Digital Signal Processing

TOPICS IN DIGITAL SIGNAL PROCESSING

Digital Signal Processing

Laboratory Experiments
Using C and the TMS320C31 DSK

RULPH CHASSAING
University of Massachusetts, Dartmouth

A Wiley-Interscience Publication
JOHN WILEY & SONS, INC.
New York • Chichester • Weinheim • Brisbane • Singapore • Toronto

Copyright © 1999 by John Wiley & Sons, Inc.

Library of Congress Cataloging in Publication Data:

Chassaing, Rulph.
    Digital signal processing : laboratory experiments using C and the
TMS320C31 DSK / by Rulph Chassaing.
       p. cm. — (Topics in digital signal processing)
    Includes bibliographical references and index.
    ISBN 0-471-29362-8 (alk. paper)
    1. Signal processing—Digital techniques—Experiments.   I. Title.
II. Series.
TK5102.9.C474   1998
621.382'2—dc21   98-7855
                                                   CIP

Printed in the United States of America

10 9 8 7 6 5 4

Contents

Preface

Digital signal processors, such as the TMS320 family of processors, are found in a wide range of applications such as in communications and controls, speech processing, and so on. They are used in Fax, modems, cellular phones, etc. These devices have also found their way into the university classroom, where they provide an economical way to introduce real-time digital signal processing (DSP) to the student.

With the introduction of Texas Instruments' third-generation TMS320C3x processor, floating-point instructions and a new architecture that supports features which facilitate the development of high-level language compilers appeared. The C optimizing compiler takes advantage of the special features of the TMS320C3x processor such as parallel instructions and delayed branches. Throughout the book, we refer to the C/C++ language as simply C. Generally, the price paid for going to a high-level language is a reduction in speed and a similar increase in the size of the executable file. Although TMS320C3x/assembly language produces fast code, problems with documentation and maintenance may exist. A compromise solution is to write time-critical routines in TMS320C3x code that can be called from C.

This book is intended primarily for senior undergraduate and first-year graduate students in electrical and computer engineering and as a tutorial for the practicing engineer. It is written with the conviction that the principles of DSP can best be learned through interaction in a laboratory setting, where the student can appreciate the concepts of DSP through real-time implementation of experiments and projects. The background assumed is a system course and some knowledge of assembly language or a high-level language such as C.

Most chapters begin with a theoretical discussion, followed by representative examples that provide the necessary background to perform the concluding experiments. There are a total of 60 solved programming examples using both TMS320C3x and C code. Several sample projects are also discussed.

Programming examples using both TMS320C3x and C code are included throughout the text. This can be useful to the reader who is familiar with both DSP and C programming, but who is not necessarily an expert in both. Although the reader who elects to study the programming examples in either TMS320C3x or C code will benefit from this book, the ideal reader is one with an appreciation for both TMS320C3x and C code.

This book can be used in the following ways:

1. For a laboratory course using many of the Examples and Experiments from Chapters 1-7. The beginning of the semester can be devoted to short programming examples and experiments and the remainder of the semester used for a final project.
2. For a senior undergraduate or first-year graduate design project course, using Chapters 1-5, selected materials from Chapters 6-8, and Appendices C and D.
3. For the practicing engineer as a tutorial and for workshops and seminars.

Chapter 1 introduces the tools through three examples. These tools include an assembler and a debugger that are provided with the DSP Starter Kit (DSK). Program examples in C can be tested without a C compiler since all associated executables files are on the accompanying disk. Chapter 2 covers the architecture and the instructions available for the TMS320C3x processor. Special instructions and assembler directives that are useful in DSP are discussed. Chapter 3 illustrates input and output (I/O) with the two-input analog interface chip (AIC) on the DSK board through several programming examples. An alternative I/O with a 16-bit stereo audio codec that can be interfaced with the DSK is described.

Chapter 4 introduces the z-transform and discusses finite impulse response (FIR) filters and the effect of window functions on these filters. Chapter 5 covers infinite impulse response (IIR) filters. Programming examples to implement FIR and IIR filters, in both TMS320C3x and C code, are included.

Chapter 6 covers the development of the fast Fourier transform (FFT). Programming examples on FFT are included. Chapter 7 demonstrates the usefulness of the adaptive filter for a number of applications with the least mean square (LMS). Chapter 8 discusses a number of DSP applications.

A disk included with this book contains all the programs discussed in the text. See page xv for a list of the programs/files included on the disk.

During the summers of 1996-1998, a total of 115 faculty members from over 100 Institutions took my DSP and Applications workshops supported by grants from the National Science Foundation (NSF). I am thankful to them for their encouragement, participation and feedback on this book. In particular, Dr. Hisham Alnajjar from the University of Hartford, Dr. Armando Barreto from Florida International University, Dr. Paul Giolma from Trinity University, Dr. William Monaghan from the College of Staten Island—CUNY, and Dr. Mark Wickert from the University of Colorado at Colorado Springs. I also thank Dr. Darrell Horning from the University of New Haven, with whom I coauthored the text *Digital Signal Processing with the TMS320C25,* for introducing me to book-writing. I thank all the students who have taken my DSP and Senior Design Project courses. I am particularly indebted to two former students, Bill Bitler and Peter Martin, who have worked with me for many years and have contributed to this book as well as to my previous book *Digital Signal Processing with C and the TMS320C30.*

The support of the National Science Foundation's Undergraduate Faculty Enhancement (UFE) Program in the Division of Undergraduate Education, Texas Instruments, and the Roger Williams University Research Foundation is appreciated.

RULPH CHASSAING

List of Examples

List of Programs/Files on Accompanying Disk

README	TXT	169
EGAVGA	BGI	5363

Directory of CH1

MATRIX	ASM	1628
SINE4P	ASM	1118
MATRIXC	ASM	6860
MATRIXC	C	482
MATRIXC	CMD	750
MATRIXC	OUT	1901
AICCOM31	ASM	5308

Directory of CH2

ADD4	ASM	702
MULT4	ASM	1150
FIR4	ASM	3016
MATRIXMF	ASM	1369
ADDMFUNC	ASM	556
ADDM	ASM	4179
FIR11	ASM	2595
ADDM	C	393
MATRIXM	C	488
ADDM	CMD	804
ADDM	OUT	1878
MATRIXM	OUT	2053
FIR11L	DAT	190
FIR11X	DAT	242

Directory of CH3

INTERR	ASM	1915
SINEALL	ASM	3093
LOOP	ASM	838
LOOPI	ASM	1076
SINE8I	ASM	1539
PRNOISE	ASM	1829
PRNOISEI	ASM	2214
VECS_DSK	ASM	222
LOOPALL	ASM	8525
LOOPC	ASM	9635
PCLOOP	EXE	212306
C31COM	ASM	3169
DAQ	EXE	250093
DAQ	ASM	9627
LOOPALL	C	2488
AICCOMC	C	2271
LOOPC	C	610
LOOPCI	C	740
C31COM	C	439
C31LOOP	C	873
LOOPALL	CMD	991
LOOPCI	CMD	1029
C31COM	CMD	905
C31LOOP	CMD	905
LOOPALL	OUT	2100
LOOPC	OUT	2146
LOOPCI	OUT	2422

C31LOOP	OUT	2856
C31COM	OUT	1664
PCCOM	CPP	1309
PCLOOP	CPP	1033
DAQ	CPP	1632
DAQ	DAT	3117
DSKLIB	LIB	143872
SYMBOLS	H	4190
DSKLIB	H	293
VECS_DSK	OBJ	427
SINEFM	ASM	2622

Directory of CH4

BP45SIM	ASM	2383
LP11SIM	ASM	2385
FIRNC	ASM	2147
FIRPRN	ASM	3550
FIRMCF	ASM	2016
FIRMC	ASM	16714
AICCOMC	C	2233
FIRDMOVE	C	1106
FIRERIC	C	1509
FIRMC	C	713
FIRC	C	1376
FIRMC	CMD	1091
FIRDMOVE	OUT	3018
FIRERIC	OUT	3113
FIRMC	OUT	3040
FIRC	OUT	3089
BP45SIM	DAT	371
LP11SIM	DAT	190
FIR	BAT	97
FIRPROGA	BAS	20237
FIRPROG	BAS	17752
BP55	COF	1080
PASS2B	COF	1083
PASS3B	COF	1088
LP55	COF	1095
BS55	COF	1082
LP11	COF	578
HP55	COF	1079
PASS4B	COF	1084
STOP3B	COF	1086
BP23	COF	551
BP41	COF	804

BP45	COF	843
BP33	COF	706
COMB14	COF	273
KBP53	COF	2426
RBP53	COF	2424
BP45COEF	H	721

Directory of CH5

SINEA	ASM	1767
COSINEA	ASM	1833
IIR6BP	ASM	2335
SINEC	C	1971
IIR6BPC	C	1057
IIR6BPC	CMD	1033
SINEC	OUT	3986
IIR6BPC	OUT	3115
AMPLIT	CPP	17889
BLT	BAS	5363
IIR6COEF	H	639
SINECMOD	C	2456
SINESW	ASM	2553

Directory of CH6

TWID128	ASM	2096
FFT_RL	OBJ	1011
FFT_RL	ASM	6358
TWID8	ASM	221
FFT128C	C	2498
FFT	C	2294
SINEGEN	C	540
TWIDGEN	C	814
FFT8C	C	680
FFT8MC	C	1124
FFT128C	CMD	1033
FFT128C	OUT	8327
FFT8C	OUT	5837
FFT8MC	OUT	2985
TWIDDLE	H	8557
COMPLEX	H	212
FFT8C	CMD	922

Directory of CH7

ADAPTP	ASM	4110
NOTCH2W	ASM	4072

ADAPTER	ASM	3848
ADAPTC	C	1684
ADAPTDMV	C	1600
ADAPTIVE	C	7783
ADAPTSH	C	1938
ADAPTTB	C	1639
ADAPTDMV	CMD	983
ADAPTSH	CMD	746
ADAPTDMV	OUT	3414
ADAPTSH	OUT	5227
ADAPTTB	OUT	4543
SIN312		694
SIN312A		776
HCOS312		686
HCOS312A		749
COS312A		798
DPLUSN		730
DPLUSNA		840
SCDAT		3985
SIN1000		647
SHIFT	C	812
ADAPTERC	ASM	4321

Directory of CH8

MR7DSK	ASM	33624
FIR8SETS	ASM	10251
FIRALL	ASM	10311
MR10SRAM	ASM	46118
ALARMGEN	ASM	6053
SIM2	C	3803
FIRALL	CPP	1226
FIR8SETP		3057
FIRALL	EXE	212589
EISINE	C	1521
EISINE	CMD	1061
EISINE	OUT	4307
VEC_DSK	ASM	215
VEC_DSK1	ASM	290
SINE4INT	C	1454

SINE4INT	CMD	947
SINE4INT	OUT	4035
SINE4C	C	959
SINE4C	CMD	959
SINE4C	OUT	2317
FIREXT	ASM	11292

Directory of APPB

BP45SIMP	ASM	2573
BP45SIMP	DAT	788
DAQ	DAT	3117
MATBP33	COF	594
MAT33	M	523
MAT63	M	544
DAQ	M	752

Directory of APPC

SINEHEX	C	1254
BP45HEX	C	1580
TESTMEM	CPP	3690
C31DLHEX	CPP	2087
SINEHEX	CMD	1015
SINHEX30	CMD	448
BP45HEX	CMD	1048
BPHEX30	CMD	471
BP45HEX	OUT	3177
BP45HEX	A0	4717
SINEHEX	OUT	2632
SINEHEX	A0	3113
SINEHEX	MAP	4439

Directory of APPD

LOOPL_CS	ASM	865
LOOPR_CS	ASM	848
LOOPB_CS	ASM	1015
CSCOM	ASM	6646
BP45CS	ASM	2702

1

Digital Signal Processing Development System

- Use of the TMS320C31 DSK
- Testing the software and hardware tools such as the debugger
- Programming examples in C and TMS320C3x code to test the tools

Chapter 1 introduces several tools available for digital signal processing (DSP). These tools include the TMS320C31-based DSP Starter Kit (DSK) with complete input and output support. Three examples are included to illustrate these development tools and, in particular, to test the DSK.

1.1 INTRODUCTION

Digital signal processors, such as the TMS320C31, are just like fast microprocessors with a specialized instruction set and architecture appropriate for signal processing. The architecture of a digital signal processor is very well suited for numerically intensive calculations. These processors are used for a wide range of applications from communications and controls to speech and image processing. They are found in music synthesizers, cellular phones, fax/modems, etc. They have become the product of choice for a number of consumer applications, since they can be very cost-effective. DSP techniques have been very successful because of the development of low-cost software and hardware support. For example, applications such as modems and speech recognition can be less expensive using DSP techniques. Furthermore, general-purpose digital signal processors can handle different tasks, since they can be readily reprogrammed for a different application. While analog-based systems with discrete electronic components such as resistors can be more sensitive to temperature changes,

1

DSP-based systems are less affected by environmental conditions such as temperature.

Books and articles have been published that address the importance of digital signal processors for a number of applications [1–17]. Various technologies have previously been used for signal processing. The more common applications using DSP processors have been for the audio-frequency range from 0 to 20 kHz, for which they have been very suitable. Speech can be sampled at 10 kHz, which implies that each sample or value is acquired at a rate of 1/(10 kHz) or 0.1 ms. For example, a commonly used sample rate (how quickly samples are acquired) of a compact disk (CD) is 44.1 kHz.

The basic system consists of an analog-to-digital converter (ADC) to capture an input signal. The resulting digital representation of the captured signal is then processed by a digital signal processor such as the TMS320C31 and then output through a digital-to-analog converter (DAC). Also included within the basic system is a special input filter for antialiasing to eliminate erroneous signals, and an output filter to smooth or reconstruct the processed output signal.

Most of the work presented here involves the design of a program to implement a DSP application.

1.2 DSK SUPPORT TOOLS

To perform the experiments, the following tools are needed:

1. Texas Instruments' DSP Starter Kit (DSK), which includes a board with the TMS320C31 floating-point processor and input and output (I/O) support. The DSK board contains an analog interface circuit (AIC) chip that provides for programmable ADC and DAC rates, and input and output filtering, all on a single chip. Software tools for assembling and debugging as well as several applications examples are also included with the DSK package [18].

2. An IBM compatible PC. The DSK board connects to the parallel printer port in the PC, through a DB25 cable provided with the DSK package.

3. An oscilloscope, signal generator, speakers, and signal/spectrum analyzer (optional). Shareware utilities are available that utilize the PC and a sound card to create a virtual instrument such as an oscilloscope, a function generator, or a spectrum analyzer (see Section 1.4 and Appendix B).

4. TMS320 floating-point DSP assembly language tools (optional) to support C programs [19–23]. These tools include a C compiler, an assembler (different than the one provided with the DSK), and a linker that creates an executable common-object file format (COFF) file that can run on the DSK [24]. They are not needed to run and test the C programs listed in

this book and included on the accompanying disk, as long as these programs are not modified.

The DSK based on the TMS320C31 (C31) is a relatively powerful, yet inexpensive ($99) development board for real-time digital signal processing. The DSK board contains the TMS320C31 processor and the TLC320C40 analog interface circuit (AIC) chip for input and output [18].

The assembler provided with the DSK creates an executable file that can be directly downloaded into the C31 on the DSK and run. It does not create a COFF file, which is obtained using the TMS320 floating-point DSP assembly language tools. The DSK assembler does not include or require a linker. Code is assembled at an absolute address into specified memory sections using certain assembler directives. These directives serve as a linker and can be used to include or chain several files together (discussed in Chapter 2). The assembled executable file can be loaded into the C31 on the DSK by using the debugger or boot loader provided with the DSK package, as illustrated later in this chapter.

1.3 PROGRAMMING EXAMPLES TO TEST THE DSK TOOLS

Three examples are introduced to illustrate the DSK tools. Don't worry about the program code at this point, since these programs are only to test the tools, in particular, the DSK. All the programs discussed in this book are on the accompanying disk. The programs coded in TMS320C3x or assembly language were assembled using the DSK software tools version 1.22. The latest version of these tools is available from Texas Instruments' FTP site at FTP.TI.COM.

1. The DSK package includes a User's Guide manual, a DB25 parallel printer cable, and a disk that contains the assembler, debugger, and various utilities and applications examples [18,19]. The DSK (board) requires a DC adapter that provides 7.5–12 Volts DC or an AC adapter that provides 6–9 Volts AC; both must supply a minimum of 400 milliamps [18]. Adapters with lower voltage or amperage specifications than recommended should not be utilized. When powered up, the light on the DSK board should change color (green and red). When the DSK is not properly connected, it is usually because of the parallel port selection. For example, the address is 0x378 for LPT1 (by default). If that port address is already being used, select another communication port (0x278 for LPT2 or 0x3BC for LPT3). RCA type connectors are available on the DSK board for input and output.

2. Create a directory `dsktools` and install the software tools provided on the disk included with the DSK package. Add `dsktools` to the path in your `autoexec.bat` file using `PATH = C:\dsktools` so that the software tools can be accessed from other directories, with C representing the selected hard drive. The number of active files should be limited to 20 when using the debugger by setting FILES = 20 in your config.sys file.

Example 1.1 Matrix/Vector Multiplication Using TMS320C3x Code

This example illustrates the use of some of the tools. Don't worry at this point about the program code. Figure 1.1 shows a listing of the MATRIX.ASM program to multiply a (3 × 3) matrix A by a (3 × 1) vector B, or

```
;MATRIX.ASM - MATRIX/VECTOR MULTIPLICATION (3x3)x(3x1) = (3x1)
          .start ".data",0x809C00   ;starting address for data
          .start ".text",0x809900   ;starting address for text
          .data                     ;data section
A         .float 1,2,3,4,5,6,7,8,9  ;values for matrix A
B         .float 1,2,3              ;values for matrix B
A_ADDR    .word  A                  ;starting address of matrix A
B_ADDR    .word  B                  ;starting address of matrix B
OUT_ADDR  .word  $                  ;output (current) address
          .entry BEGIN              ;start of code
          .text                     ;text section
BEGIN     LDP    A_ADDR             ;init to data page 128
          LDI    @A_ADDR,AR0        ;AR0=starting address of A
          LDI    @B_ADDR,AR1        ;AR1=starting address of B
          LDI    @OUT_ADDR,AR2      ;AR2= output address
          LDI    3,R4               ;R4 used as LOOPI counter
LOOPI     LDF    0,R0               ;initialize R0=0
          LDI    2,AR4              ;AR4 used as LOOPJ counter
LOOPJ     MPYF3  *AR0++,*AR1++,R1   ;R1=A[I,J]*B[J]
          ADDF3  R1,R0,R0           ;accumulate in R0
          DB     AR4,LOOPJ          ;decrement AR4.Branch until AR4<0
          FIX    R0,R2              ;convert R0 from float to integer
          STI    R2,*AR2++          ;store integer output in memory
          LDI    @B_ADDR,AR1        ;AR1=starting address of matrix B
          SUBI   1,R4               ;decrement R4
          BNZ    LOOPI              ;branch while R4 is not zero
          BR     $                  ;branch to current addr (itself)
```

FIGURE 1.1   Matrix/vector multiplication program using TMS320C3x code (MATRIX.ASM).

$$\begin{bmatrix} 1 & 2 & 3 \\ 4 & 5 & 6 \\ 7 & 8 & 9 \end{bmatrix} \begin{bmatrix} 1 \\ 2 \\ 3 \end{bmatrix} = \begin{bmatrix} 14 \\ 32 \\ 50 \end{bmatrix}$$

that yields a (3×1) vector containing the result (14, 32, 50). All the programs discussed in this book are included on the accompanying disk.

Assembling

Assemble the source program MATRIX.ASM by typing:

```
dsk3a matrix.asm
```

The asm extension is not necessary, but it is a good practice to name the source file with an extension asm. The assembler creates the executable file MA-TRIX.DSK (not case-sensitive) that can be downloaded into the C31 on the DSK and run using either the debugger or the boot loader (boot loading is illustrated in Example 1.2).

Loading an Executable File Into the DSK to Run

To invoke the debugger, type:

```
dsk3d
```

If the debugger is not successfully invoked, check for proper power supply connection and the parallel printer port setup in your PC. The DSK connection to the parallel port on the PC defaults to LPT1. [18].

The C31 processor should always be reset before running a program. Within the debugger, you can reset the C31 with the command:

```
reset
```

Then load the executable file MATRIX.DSK by typing the command:

```
load matrix.dsk
```

These commands are not case-sensitive and the extension dsk is not necessary. The debugger screen should now look as in Figure 1.2.

Note that the program code starts at the memory location 809900, as shown in the first column within the **DISASSEMBLY** window. The hexadecimal notation 0x is implied. The first column represents the instruction memory address and the second column represents the instruction opcode.

1. Press F8 to single-step through the first five lines of code in the matrix program, shown in Figure 1.2. A summary of the instructions available for the

```
──────── DISASSEMBLY ────────            ────CPU REGISTERS────
809900 .text    LDIU 00080h,DP           PC   0080990b SP   00809eff
809901          LDI @A_ADDR,AR0          F0 0360000000 F1 0310000000
809902          LDI @B_ADDR,AR1          F2 000000000e F3 0000000000
809903          LDI @OUT_ADDR,AR2        F4 0000000003 F5 0000000000
809904          LDI 3,R4                 F6 0000000000 F7 0000000000
809905 LOOPI    LDF 0.000000e+00,R0      AR0  00809c03 AR1  00809c0c
809906          LDI 2,AR4                AR2  00809c0e AR3  000000ee
809907 LOOPJ    MPYF3 *AR0++(1),*AR1++(1),R1   AR4  00ffffff AR5  00000000
809908          ADDF3 R1,R0,R0           AR6  00000000 AR7  00000000
809909          DB AR4, LOOPJ            IR0  00000002 IR1  00000000
80990a          FIX R0,R2                ST   00000000 RC   00000000
80990b          STI R2,*AR2++(1)         RS   000000ee RE   000000fd
80990c          LDI @B_ADDR,AR1          DP   00000080 BK   00000000
80990d          SUBI 1,R4                IE   000000c4 IF   00000300
80990e          BNE LOOPI                IOF  00000000 _dT  00000002
```

```
──── COMMAND ────               ──── MEMORY ────
  reset               809800   ea9b9e0f 00000063 f6eb7eef f55a7fff
  load matrix         809804   ffbdffff 5ffdcfff 7ad7ff7d fb7df67f
> ■                   809808   6fd7ffff f7d8fffb ee7fdfbf dffec7ff
                      80980c   fd7eadbb ff8fff8a ff7d7fff dfcdeff5
                      809810   fe7fffdf fdffbbdf f7fdf7fa 8deafebc
                      809814   fffdf6db bff3fffb f5fff79f fffff7f7
```

F1Help F2REG40 F3FLOAT F4Srce F5Run F6DispBP F7ClrAll F8SStep F9Grow F10FStep

FIGURE 1.2 DSK Debugger window screen.

C31 is listed in Appendix A; however, don't worry for now about the code. The right-top window screen shows that AR0 contains the value 809c00, which is the starting address in memory where the nine values of the matrix A are stored. AR1 contains the value 809c09, the starting address in memory where the three values of the vector B are stored; and AR2 contains 809c0e, the starting memory address for the three resulting output values. The matrix A multiplied by the vector B yields the values e, 20, 32 in hex, equivalent to the decimal values 14, 32, 50.

In Chapter 2, we will see that AR0–AR7 are eight registers on the C31 that are often used to designate a specific address in memory that contains an instruction or a data value. F0–F7 represent the eight registers R0–R7 that are often used to contain a data value. The C31 has 2K words (32-bit) of internal or on-chip memory (16 million memory spaces total); 809800 (in hex) represents the starting address of this block of internal memory.

2. Access the **MEMORY** window using ALT-M (the ALT key together with M). Use the down-arrow key to scroll down from 809800, the memory address in the first column, to 809c00. Or, press ESC to access the command window and type mem 0x809c00 to display the contents in memory starting at the address 809c00. The notation 0x is necessary with a command. Press F1 for help on available debugger commands. The data values stored starting at memory address 809c00 are in floating-point format which you need not to worry about. The floating-point value 00000000 corresponds to a decimal value of 1.

Type memf 0x809c00 to display the content in memory in float (decimal) format starting at the address 809c00. Verify that the nine values of the matrix *A* are stored in memory starting at the address 809c00, followed by the three values of the vector *B*, starting at the address 809c09.

3. As you single-step through the program and execute each time the instruction STI R2,*AR2++(1), observe the register window within the debugger (top-right window screen), which displays the contents of the CPU registers in hexadecimal format, by default. Press F3 and verify that each resulting value 14, 32, 50 is contained in F0, which represents the register R0. Also, verify from the **MEMORY** window that the resulting output values e, 20, 32 in hex are stored in memory starting at the address 0x809c0e, specified in AR2. The three resulting values in memory locations 809c0e, 809c0f, and 809c10 can be displayed in 32-bit hex format or in 32-bit signed format (decimal) with the commands memx 0x809c0e or memd 0x809c0e, respectively. While a debugger command is not case-sensitive, the 0x notation for hexadecimal is required.

Press F2 or F3 to display the CPU registers in 32-bit hex format or float format, respectively. The F's represent the float of the extended precision registers R0-R7. These registers are on the C31 and displayed as F0-F7 within the CPU registers window screen.

4. Run the program again by typing reset and load matrix from the **COMMAND** window, and press F5 to run. Then, press ESC to stop execution, since the instruction BR WAIT to branch back to itself (to wait) is still being executed continuously. Note that F0 = 50, the last result.

5. Reset and load again the matrix program. Press ALT-D to access the **DISASSEMBLY** window. Use the down-arrow key to scroll down to the STI R2,*AR2++(1) instruction at the address 80990b. Press F2 to toggle or set a breakpoint, which will highlight the instruction set with the breakpoint. Press F4 to run until breakpoint. Note that the program counter (PC) contains 80990b, the address of the instruction to be executed next. Press F8 once to execute that instruction. Verify from the **MEMORY** window that the content in memory location 809c0e is the first resulting value of 14 (e in hex). Press F4 again to run until the set breakpoint, then F8 to execute the instruction STI R2,*AR2++(1) a second time, and verify the second resulting value 32 in memory location 809c0f. Repeat this process a third time to verify the third resulting value of 50 in memory location 809c10.

We will see in Chapter 2 that the instruction STI R2,*AR2++(1) stores each result from R2 into a memory location specified by AR2. The register AR2 is incremented for each output value and contains the address in memory where each result is stored. In this fashion, AR2 is used as a "pointer," pointing to a memory address. Type quit from the command window to exit the debugger.

Example 1.2 Sine Generation with Four Points Using TMS320C3x Code

This example illustrates the generation of a sinusoid using a table look-up method. There are two RCA connectors next to the light on the DSK board, one for input and the other for output. Connect the DSK output to a speaker to hear a generated tone or to an oscilloscope to view the generated sinusoidal waveform. An analog interface circuit (AIC) chip, on board the DSK, provides I/O capabilities and will be discussed in Chapter 3.

Section 1.4 and Appendix B describe several tools available as virtual instruments that can utilize the PC and a sound card as an oscilloscope or as a spectrum analyzer. For example, while the C31 on the DSK is running, the shareware utility Goldwave can be accessed and run as an oscilloscope to verify the generated output sinusoid. The output of the DSK would then be connected to the input of a sound card (such as Sound Blaster) plugged on a PC.

Figure 1.3 shows the program listing SINE4P.ASM, which generates a tone using four points. Again, don't worry about the code for now, since the emphasis is to become more familiar with the tools. This program invokes (includes) another program AICCOM31.ASM (on the accompanying disk), which contains several routines for communication with the on-board AIC for real-time input

```
;SINE4P.ASM - GENERATES A SINE USING ONLY 4 POINTS
          .start    ".text",0x809900    ;starting address for text
          .start    ".data",0x809C00    ;start address for data
          .include  "AICCOM31.ASM"      ;AIC communication routines
          .data                         ;data section
AICSEC    .word     162Ch,1h,4892h,67h  ;Fs = 8 kHz
SINE_ADDR .word     SINE_VAL            ;address of sine values
          .brstart  "SINE_BUFF",16      ;align sine table
SINE_VAL  .word     0,1000,0,-1000      ;sine values
LENGTH    .set      4                   ;length of circular buffer
          .entry    BEGIN               ;start of code
          .text                         ;text section
BEGIN     CALL      AICSET              ;initialize AIC
          LDI       LENGTH,BK           ;BK = size of buffer
          LDI       @SINE_ADDR,AR1      ;AR1 = addr of sine values
LOOP      LDI       *AR1++%,R7          ;R7 = table value
          CALL      AICIO_P             ;call AICIO for output
          BR        LOOP                ;loop back
          .end                          ;end
```

FIGURE 1.3 Sine generation program using TMS320C3x code (SINE4P.ASM).

and output capabilities. While we will discuss the AIC in Chapter 3, we will mostly use the AIC communication routines by simply "including" the file AICCOM31.ASM in other programs (fourth line in SINE4P.ASM).

1. Assemble the program SINE4P.ASM only and not AICCOM31.ASM.

2. Access the debugger, reset the C31 processor as in Example 1.1, and load the program SINE4P.DSK.

3. Press F5 to run and verify a tone with a frequency of 2 kHz. The frequency f of the resulting output waveform is obtained using:

$$f = F_s/(\text{number of points})$$

where $F_s = 8$ kHz is the sampling frequency, which also designates the output sample rate. This rate determines how fast an output sample point representing the generated sinusoidal waveform is produced. The sampling rate is specified by the A/D and D/A converters on the AIC. Although there is no external input, an output sample point is generated every $T = 1/F_s = 0.125$ ms, where T represents the sampling period.

Loading and Executing Using the Boot Loader

Run the sine generation program by invoking a boot loader program provided with the DSK software tools. This procedure does not access the debugger. Type

```
dsk3load sine4p.dsk
```

to load and run this program. Verify that a 2-kHz signal is generated. Again the extension dsk is not necessary. Care must be exercised when running a program with the boot loader, since it does *not* reset the C31. Erroneous values can result, for example, if an interrupt-driven program (interrupt will be discussed in Chapter 3) was previously loaded into the C31. In such cases, use the debugger to reset the C31.

Don't modify the original programs on the accompanying disk. Before making any changes to any file on the accompanying disk, copy it first into your hard drive.

1. **Changing the number of points to change the generated output frequency. a)** Replace the four points specified in the program (9th line) with the following eight points:

$$0, 707, 1000, 707, 0, -707, -1000, -707$$

that represent a sequence of eight points from a sinusoid taken every 45 degrees and scaled. Change also the length (LENGTH) from 4 to 8. Rename this program SINE8P.ASM. Reassemble SINE8P.ASM only and run it using the debugger or the boot loader. The file AICCOM31.ASM is included in

SINE8P.ASM and should not be assembled separately. Verify a generated sinusoidal tone with a lower pitch or frequency, $f = 8,000/8 = 1$ kHz.

b) Replace the eight-point sequence with 12 points taken every 30 degrees from a sinusoid, i.e., 0, 500, 866, ... , –500, and scaled. Change the length to 12. Verify a generated output sinusoidal tone with a frequency of $f = 8,000/12 = 666.66$ Hz.

2. Changing the sampling frequency F_s. Four values are defined/set in AICSEC (6th line in SINE4P.ASM). The first and third values specify the AIC sampling frequency F_s. Change these values such that AICSEC is set to:

$$0E1Ch, \quad 1h, \quad 3872h, \quad 67h$$

which specifies a sampling frequency $F_s = 16$ kHz, as will be shown in Chapter 3. These four values are specified in hex with an h after each value (or 0x before the value). Reassemble the program SINE4P.ASM and use the boot loader to load and run this program. Verify that the frequency of the new generated sinusoid is 4 kHz, since

$$f = 16,000/(\text{number of points})$$

3. Changing the AIC master clock to change F_s. The first and the third value specified in AICSEC are calculated in Chapter 3 using a specific value for the AIC master clock. Changing the master clock frequency proportionately changes the sampling frequency F_s.

a) Back up the file AICCOM31.ASM (on disk) and change the instruction (twelfth line in the program):

$$LDI \ 1, R0$$

to LDI 0,R0, which doubles the AIC master clock and effectively doubles the sampling frequency F_s with the values specified in AICSEC. Reassemble the original program SINE4P.ASM with a four-point look-up table and a frequency set for 8 kHz. Note that the file AICCOM31.ASM should *not* be assembled separately. Since it is "included" or incorporated in the program SINE4P.ASM, only that program is to be assembled. Use the boot loader to run the resulting executable file SINE4P.DSK. Verify that the generated output signal has a frequency of 4 kHz since the new sampling frequency is 16 kHz, or

$$f = 16,000/4 = 4 \text{ kHz}$$

b) The instruction LDI k,R0 with k = 2, 3, 4, ... , can be used to divide the AIC master clock. Let k = 2 and reassemble the original sine generation program SINE4P.ASM and verify an output signal with a frequency of 1 kHz, since F_s is effectively reduced from 8 kHz to 4 kHz. Let k = 4, and verify that the generated output signal frequency is

$$f = (8,000/\text{k})/4 = 500 \text{ Hz}$$

Example 1.3 Matrix/Vector Multiplication Using C code

You can test and run all the C-program examples in this book, since all the resulting executable files, compiled/assembled and linked with the TMS320 floating-point assembly language tools, are included on the accompanying disk. However, if the C-source file is modified, it will need to be compiled, assembled, and linked again.

Running C Programs Without the Floating-Point Tools

The source program MATRIXC.C listed in Figure 1.4 is the C version of the program MATRIX.ASM in Example 1.1. Access the debugger as in Example 1.1. Reset the C31 by typing the debugger command:

```
reset
```

```c
/*MATRIXC.C - MATRIX/VECTOR MULTIPLICATION */
main()
{
 volatile int *IO_OUTPUT = (volatile int *) 0x809802;
 float A[3][3] = { {1,2,3},
                    {4,5,6},
                    {7,8,9} };
 float  B[3] = {1,2,3};
 float result;
 int i,j;
 for (i = 0; i < 3; i++)
  {
  result = 0;
  for (j = 0; j < 3; j++)
   {
    result += A[i][j] * B[j];
   }
  *IO_OUTPUT++=(int)result; /*result start in mem addr 0x809802*/
  }
}
```

FIGURE 1.4 Matrix/vector multiplication program using C code (MATRIXC.C).

Then, within the debugger type the command

```
load matrixc.out
```

to load the executable COFF file MATRIXC.OUT (not case-sensitive) supplied on the accompanying disk. The extension OUT is not necessary, since the debugger detects such type of executable COFF file as opposed to an executable file with a dsk extension.

Single-step through the program up to the instruction STI RS,*AR0 at the memory address 80983e. Note that there is much initialization code added from compiling. The STI instruction causes each resulting output value to be stored in consecutive memory, starting at the address pointed by AR0, which contains the output address 809802. Verify the three resulting values e, 20, 32 in hex. Type memd 0x809802 to verify from the memory-window screen the three resulting values 14, 32, and 50 stored in memory addresses 809802-809804.

The DSK does not support a C-source debugger. Hence, the C-source file cannot be displayed through the DSK debugger window screen. With a C-source debugger, one could single-step through an instruction in C and observe the corresponding steps through equivalent assembly instructions [1, 23]. Tools that support debugging capabilities, such as the C3x debugger for the evaluation module (EVM), are available from Texas Instruments [25], and Code Composer is available from GO DSP (see Section 1.4)

C Compiling and Linking Using Floating-Point Tools

a) Compiling/Assembling. This section illustrates the use of the TMS320 floating-point DSP assembly language tools, version 5.0 [21–23]. These tools are not included with the DSK package. The C-code programs in this book were compiled/assembled and linked with these tools. Compile/assemble the C-source program MATRIXC.C, by typing:

```
cl30 -k matrixc.c
```

The extension c is not necessary. This creates the source file MATRIXC.ASM as well as the object file MATRIXC.OBJ. Various compiler options are available [22]. The -k option is to retain the assembly source file MATRIXC.ASM, since the CL30 command compiles and assembles. Different levels of optimization are available for compiling. Using CL30 -o3 selects the highest optimization level (register, local, global, and file) for faster execution speed. The -o2 option invokes the second level (by default) of optimization (without the file optimization available with the -o3 option). Care must be exercised when invoking these optimization levels if the resulting executable files is to be downloaded and run on the DSK, especially with older versions of the DSK tools.

A source file in TMS320C3x assembly code such as MATRIX.ASM can be assembled with the floating-point tools using the command

ASM30 MATRIX.ASM

to create the object file MATRIX.OBJ. Note that the command CL30 MATRIXC.C compiles and assembles in one step.

b) Linking. Link the resulting object file MATRIXC.OBJ using the sample linker command file MATRIXC.CMD listed in Figure 1.5 (on the accompanying disk), by typing:

lnk30 matrixc.cmd

This creates the executable file MATRIXC.OUT. This is a linked common-object file format (COFF), popular in Unix-based systems and adopted by several makers of digital signal processors. The COFF format makes it easier for modular programming and managing code segments [24].

Note that the comments /* and */ used in C programming have the same functions in the linker command file MATRIXC.CMD shown in Figure 1.5. The

```
/*MATRIXC.CMD - LINKER COMMAND FILE               */
-c                      /*using C convention      */
-stack 0x100            /*256 words stack         */
matrixc.obj             /*object file             */
-O matrixc.out          /*executable output file  */
-l rts30.lib            /*run-time library support*/
MEMORY
 {
  RAMS: org=0x809800, len=0x2   /*boot stack      */
  RAM0: org=0x809802, len=0x3FE /*internal block 0*/
  RAM1: org=0x809C00, len=0x3C0 /*internal block 1*/
 }
SECTIONS
 {
  .text:   {} > RAM0        /*code               */
  .cinit:  {} > RAM0        /*initialization tables*/
  .stack:  {} > RAM1        /*system stack        */
 }
```

FIGURE 1.5 Linker command file for C coded matrix example (MATRIXC.CMD).

-1 option invokes the file RTS30.LIB included with the floating-point tools, which is an object-library file that contains run-time support C functions. Don't worry for now about the MEMORY and SECTIONS specifications within the linker command file.

1.4 ADDITIONAL SUPPORT TOOLS

The following tools can be useful in conjunction with the DSK (see also Appendix B).

1. Code Explorer is a free, scaled-down version of the popular debugger Code Composer, available from GO DSP [26]. It can be retrieved at the web site address www.go-dsp.com. as a zipped file and pkunzipped. An executable dsk file can be readily downloaded into the Code Explorer and run. A sequence of data stored within consecutive memory locations can be plotted in both the time and frequency domains within the Code Explorer debugger environment and saved on disk. The debugger includes capabilities to single-step, run to breakpoint, and modify memory/register (see Appendix B). An example on filtering is described in Appendix B to illustrate the use of the Code Explorer as a debugger for running a program and plotting the resulting output within the debugger environment. The programs in Examples 1.1 and 1.2 can be tested with the Code Explorer debugger.

Code Explorer does not support COFF executable files. Code Composer, with appropriate documentation, can be purchased from GO DSP and allows you to download and execute COFF or DSK files.

2. Goldwave, a shareware virtual instrument (goldwave.zip), can be used as an oscilloscope or as a spectrum analyzer in conjunction with a PC and a sound card such as Sound Blaster [27]. Goldwave can also be used to generate functions such as a sinusoidal signal with a specified frequency or random noise (see Appendix B). It can be retrieved at the web address www.goldwave.com.

3. DigiFilter is a filter-design package that supports the DSK and is available from MultiDSP at multidsp@aol.com. It is illustrated in Appendix B in conjunction with filtering, discussed in Chapters 4 and 5. The designed filter characteristics can be downloaded directly into the DSK and run to implement a filter in real time [28].

4. Virtual Bench is a virtual instrument available from National Instruments (which produces LabView) at www.natinst.com. With a data acquisition card that plugs onto a PC slot and an I/O board for input and output, Virtual Bench can be used as a function generator, as an oscilloscope, or as a spectrum analyzer.

5. External and Flash memory. Appendix C describes a daughter board that contains 32K words of external SRAM memory and 32K words of flash memory. The flash memory allows you to store a specific application program in the flash memory section and run on the DSK without any connection to a PC. This daughter board connects directly to the DSK through the four 32-pin connectors along the edge of the DSK board. All the TMS320C31 signals are routed to these four expansion connectors on the DSK and are available for the optional use of daughter boards with external memory or with alternative I/O capability, as described in Chapter 3 and Appendices C and D.

6. Input/Output Alternative with 16-bit Stereo Codec. Appendix D describes a board that interfaces to the DSK and contains Crystal's CS4216 (or CS4218) 16-bit stereo audio codec with two complete channels for input and output. An evaluation board based on the CS4216 (or CS4218) codec is available from Crystal Semiconductors.

7. SigLab is a virtual lab (box) with support software, available from DSPTechnology at siglab@dspt.com. The SigLab box is interfaced to a PC via an SCSI connector. A two-channel, 20-kHz bandwidth and a four-channel with a 50-kHz bandwidth are available. The SigLab box includes a TMS320C31 for real-time signal processing and two fixed-point digital signal processors from Analog Devices for filtering support. SigLab, while connected to the PC through an SCSI interface, can be accessed for real-time input generation and output monitoring while the DSK is also running. For example, it can be used as an oscilloscope or as a spectrum analyzer through one channel on the SigLab box connected to the output on the DSK, while generating signals such as a two-tone sinusoid or random noise through another SigLab channel connected to the input on the DSK.

8. RIDE40, available from Hyperception at info@hyperception.com is a virtual design tool that can be used to implement DSP algorithms. It contains a wide range of functional blocks for FFT, correlation, filtering, etc., and can be used for both simulation and real time. Within a few minutes, one can design and test a DSP system that includes functional blocks such as sine generators, filters, and the FFT. Results can be displayed on the PC monitor or to an external device such as an oscilloscope. However, it currently supports the C30-based EVM but not the C31-based DSK.

Digital filters can be readily designed with a filter package available from Hyperception.

9. Updated DSK and C3x Tools. Texas Instruments' web site contains the most recent version of the C31 DSK software tools. These tools include the assembler and debugger as well as several support and applications examples. The DSK software tools, version 1.22, were used to assemble the programs discussed in this book. Texas Instruments' FTP site is: FTP.TI.com. Select C3xdsktools to retrieve the updated software support tools for the C31 DSK.

Several applications examples are included with the DSK software package (first, assemble the support source files with the asm extension) such as:

a. DSK_OSC.EXE to use the DSK and the PC monitor as an oscilloscope.

b. DSK_SG.EXE to obtain a signal generator with the following functions: sine (SINE_SG.ASM), ramp (RAMP_SG.ASM), random (RAND_SG.ASM), and sawtooth (SAWT_SG.ASM). Test the sine generator and verify that several sinusoidal signals with different frequencies can be added and the resulting waveforms generated.

c. DSK_WAV.EXE calls DSKWAV files. Speech can be recorded as input to the DSK, then played back.

Application programs on the fast Fourier transform (FFT), discussed in Chapter 6, are also included with the DSK package. For example, FFT_512.EXE implements a 512-point FFT.

1.5 EXPERIMENT 1: TESTING THE DSK TOOLS

This experiment illustrates the use of the tools, in particular, the software and hardware support tools associated with the DSK.

1. Perform/implement the matrix program MATRIX.ASM in Example 1.1. This assembly program executes faster than its C-coded counterpart MA-TRIXC.C, discussed in Example 1.3, even though it is longer and looks more difficult.

2. Perform/implement the sine generator program in Example 1.2.

3. Perform/implement the C-code matrix program discussed in Example 1.3. Note that both the source file MATRIXC.C as well as the executable file MATRIXC.OUT are included on the accompanying disk. It is not necessary to have the TMS320 floating-point DSP assembly language tools in order to run the C programs included in this book. However, if you modify a C program, then you need these tools in order to recompile it and relink to create an executable file that can be run on the DSK. The C compiler command CL30 MA-TRIXC.C compiles and assembles to create both the TMS320C3x assembly source code MATRIXC.ASM and the object file MATRIXC.OBJ. This object file is linked with a run-time library support file RTS30.LIB to create the executable COFF file MATRIXC.OUT that can be loaded directly into the DSK and run.

4. Echo program. Run the program LOOP.ASM on the accompanying disk, referred to as a "loop" or echo program. To test this program, connect to the input of the DSK a sinusoidal signal from a function generator with an amplitude of approximately 1–3 V and a frequency between 1 and 3 kHz. Observe a delayed output sinusoidal signal of the same frequency. Vary the input frequency between 1 and 3 kHz and verify the same change in the frequency of the output signal. The values set from AICSEC in the loop program LOOP.ASM specify a sampling frequency of F_s = 8 kHz and a bandwidth of approximately 3,550 Hz (discussed in Chapter 3). This bandwidth represents the cutoff frequency of an internal input filter, on-chip the AIC, called antialiasing filter. Increase the input signal frequency above this bandwidth and verify that it is attenuated or cut-off by the internal input filter on-chip the AIC. Deleting this input filter will cause aliased output signals which we will verify in Chapter 3 (Example 3.3).

5. Test some of the applications examples that are provided with the DSK package.

REFERENCES

1. R. Chassaing, *Digital Signal Processing with C and the TMS320C30,* Wiley, New York, 1992.

2. R. Chassaing and D. W. Horning, *Digital Signal Processing with the TMS320C25,* Wiley, New York, 1990.

3. P. Papamichalis ed., *Digital Signal Processing Applications with the TMS320 Family: Theory, Algorithms, and Implementations,* Texas Instruments, Inc., Dallas, TX, Vol. 3, 1990.

4. K. S. Lin ed., *Digital Signal Processing Applications with the TMS320 Family: Theory, Algorithms, and Implementations,* Prentice Hall, Englewood Cliffs, NJ, Vol. 1, 1988.

5. R. Chassaing, "Applications in digital signal processing with the TMS320 digital signal processor in an undergraduate laboratory," in Proceedings of the 1987 ASEE Annual Conference, June 1987.

6. P. Lapsley, J. Bier, A. Shoham, and E. Lee, *DSP Processor Fundamentals Architectures and Features,* Berkeley Design Technology, 1996.

7. R. Chassaing, W. Anakwa, and A. Richardson, "Real-Time Digital Signal Processing in Education," in *Proceedings of the 1993 International Conference on Acoustics, Speech and Signal Processing (ICASSP),* April 1993.

8. R. Chassaing and B. Bitler (contributors), "Signal Processing Chips and Applications," *The Electrical Engineering Handbook,* CRC Press, Boca Raton, FL, 1997.

9. S. A. Tretter, *Communication System Design Using DSP Algorithms,* Plenum Press, New York, 1995.

10. R. M. Piedra and A. Fritsh, "Digital Signal Processing Comes of Age,"in *IEEE Spectrum,* May 1996.

11. Y. Dote, *Servo Motor and Motion Control Using Digital Signal Processors,* Prentice Hall, Englewood Cliffs, NJ, 1990.

12. I. Ahmed, ed., *Digital Control Applications with the TMS320 Family,* Texas Instruments, Inc., Dallas, TX, 1991.

13. A. Bateman and W. Yates, *Digital Signal Processing Design,* Computer Science Press, New York, 1991.

14. R. Chassaing, "The Need for a Laboratory Component in DSP Education—A Personal Glimpse," *Digital Signal Processing,* Academic Press, Jan. 1993.

15. C. Marven and G. Ewers, *A Simple Approach to Digital Signal Processing,* Wiley, New York, 1996.

16. J. M. Rabaey ed., "VLSI Design and Implementation Fuels the Signal-Processing Revolution," *IEEE Signal Processing Magazine,* Jan. 1998.

17. S. H. Leibson, DSP Development Software, *EDN Magazine,* Nov. 8, 1990.

18. *TMS320C3x DSP Starter Kit User's Guide,* Texas Instruments, Inc., Dallas, TX, 1996.

19. *TMS320C3x User's Guide,* Texas Instruments, Inc., Dallas, TX, 1997.

20. *TMS320C3x General-Purpose Applications User's Guide,* Texas Instruments, Inc., Dallas, TX, 1998.

21. *TMS320C3x/C4x Assembly Language Tools User's Guide,* Texas Instruments, Inc., Dallas, TX, 1997.

22. *TMS320C3x/C4x Optimizing C Compiler User's Guide,* Texas Instruments, Inc., Dallas, TX, 1997.

23. *TMS320C3x C source Debugger User's Guide,* Texas Instruments, Inc., Dallas, TX, 1993.

24. G. R. Gircys, *Understanding and Using COFF,* O'Reilly & Assoc., Inc., Newton, MA, 1988.

25. *TMS320C30 Evaluation Module Technical Reference,* Texas Instruments, Inc., Dallas, TX, 1990.

26. Code Explorer, from GO DSP, at www.go-dsp.com

27. Goldwave, at www.goldwave.com

28. DigiFilter, from MultiDSP, at multidsp@aol.com

29. B. W. Kernigan and D. M. Ritchie, *The C Programming Language,* Prentice Hall, Englewood Cliffs, NJ, 1988.

2

Architecture and Instruction Set of the TMS320C3x Processor

- Architecture and Instruction set of the TMS320C3x processor
- Memory addressing modes
- Assembler directives
- Programming examples using TMS320C3x assembly code, C code, and C-callable TMS320C3x assembly function.

Several programming examples included in this chapter illustrate the architecture, the assembler directives, and the instruction set of the TMS320C3x processor and associated tools.

2.1 INTRODUCTION

Texas Instruments, Inc. introduced the first-generation TMS32010 digital signal processor in 1982, the second-generation TMS32020 in 1985 followed by the C-MOS version TMS320C25 in 1986 [1–5], and the TMS320C50 in 1991. The first-generation processor contains 144×16 bits of internal or on-chip memory (RAM), with a 200-ns instruction cycle time. Most of the instructions can be executed in one instruction cycle. Members of the first-generation of processors are currently available in C-MOS versions with faster execution speeds.

The second-generation TMS320C25 contains 544×16 bits of on-chip RAM, is upward code-compatible with the TMS320C10 (C1x) family of processors, and has an instruction cycle time of 100 ns, making it capable of executing 10 million instructions per second (MIPS). Other members of the second-generation (C2x) family of processors are currently available with a faster execution speed. The TMS320C50 processor is code-compatible with the first two generations of C1x and C2x processors. Within the same generation, several versions of each of these processors—C1x, C2x, and C5x—are available with different features, such as a faster execution speed and availability of on-chip

ROM. The C1x, C2x, and C5x are fixed-point processors based on a modified Harvard architecture with separate memory spaces for data and instructions that allow concurrent accesses.

Quantization error or round-off noise from an ADC is a concern with a fixed-point processor. An A/D only uses a best estimate digital value to represent an input. For example, consider an A/D with a word length of 8 bits and an input range of ± 1.5 volts. The steps represented by the A/D are: (input range)/(2^8) = 3/256 = 11.72 mv. This produces errors which can be up to $\pm(11.72$ mv)/2 = ± 5.86 mv. Only a best estimate can be used by the A/D to represent input values that are not multiples of 11.72 mv. With an 8-bit ADC, 2^8 or 256 different levels can represent the input signal. An A/D with a larger word length such as a 16-bit A/D (currently quite common) can reduce the quantization error, yielding a higher resolution. The more bits an ADC has, the better it can represent an input signal.

The TMS320C62 (C62) is the most recent fixed-point processor, announced in 1997. Unlike the previous fixed-point processors, it is based on a very-long-instruction-word (VLIW) architecture, and is not code compatible with the previous generations of fixed-point processors. The "fixed-point" TMS320C80 processor was available before the C62 and contains four fixed-point processors and one reduced-instruction set (RISC) processor. The C62 is primarily intended for high-end applications such as video and multimedia. The floating-point TMS320C67, code compatible with the C62, was also announced in 1997; it is another member of the C6x family based on the VLIW architecture.

The TMS320C31 (C31), a general-purpose digital signal processor, is a member of the third-generation family of floating-point processors, TMS320C3x [6–10]. With a 40-ns instruction cycle time, it provides capabilities for 50 million floating-point operations per second (MFLOPS) or 25 million instructions per second (MIPS). The instruction cycle time or MIPS alone do not provide the entire measure of performance, since one needs to consider as well the efficient use of memory and the type of suitable instructions. The TMS320C31 is a true 32-bit processor capable of performing floating-point, integer, and logical operations. It contains 2K words of internal or on-chip memory and has a 24-bit address bus, making it capable of addressing 2^{24} or 16 million words (32-bit) of memory space for program, data, and input/output. With such features and special addressing modes, the C31 is very well suited for applications ranging from communication and control to instrumentation, speech, and image processing.

Even though the TMS320C31 has only one serial port whereas the TMS320C30 has two, the C31 has a faster execution speed. Connectors available on the C31 DSK serve the function of a serial port, and can be used to interface to another board with external memory or with alternative input/output capability for faster processing, as described in Appendices C and D. An application-specific integrated circuit (ASIC) has a "DSP core" with customized circuitry for a specific application. The C31 can be used as a standard general-purpose processor programmed for a specific application.

The TMS320C32 is another member of the third-generation of floating-point processors, but with one-fourth of the internal or on-chip memory available on the C31 (although it has special features for accessing external memory).

The TMS320C40 is a fourth-generation floating-point processor, code-compatible with the C3x processor. It has the same amount of on-chip memory as the C31, and six serial ports (the smaller C44 version has four serial ports). A C40 can connect directly to six other C40 processors without any glue logic, making the C40 suitable for parallel processing [11].

A fixed-point processor is better for devices such as cellular phones that use batteries, since it uses less power than an equivalent floating-point processor. The fixed-point processors C1x, C2x, and C5x have limited dynamic range and precision, whereas the floating-point processors C3x and C4x provide greater dynamic range. In a fixed-point processor, it is necessary to scale the data to reduce overflow, and this must be done with care. Overflow occurs when an operation such as the addition of two numbers produces a result with more bits than can fit within a processor's register. The 40-bit extended precision registers R0–R7 available on the TMS320C3x make it possible to accumulate without risking overflow. These registers are 40 bits wide, even though the busses on the C31 are 32 bits wide. These extra bits provide more accuracy while avoiding overflow. The floating-point representation used by Texas Instruments is not the standard IEEE 754 floating-point format for data representation. Although a floating-point processor is generally more expensive, since it has more "real estate" or is a larger chip because of additional circuitry, it is generally easier to program; and floating-point support tools are easier to use. The fixed-point C compiler available for the C1x, C2x, and C5x fixed-point processors is not as efficient as the floating-point C compiler that supports the C3x/C4x processors. A fixed-point type is not included in the ANSI C standard, whereas a floating-point compiler can take advantage of the floating-point hardware.

Other digital signal processors are available, such as the DSP96000 from Motorola Inc.and the ADSP21060 SHARC [12] from Analog Devices Inc.

2.2 TMS320C3x ARCHITECTURE AND MEMORY ORGANIZATION

The TMS320C31 has 2K words (32-bit) of internal or on-chip memory and 2^{24} or 16 million words of addressable memory containing program, data, and input/output space. In a von Neumann architecture, program instructions and data are stored in a single memory space. A processor with a von Neumann architecture can make a read or a write to memory during each instruction cycle. Typical DSP applications require several accesses to memory within one instruction cycle.

The TMS320C3x is based on a modified Harvard architecture, with independent memory banks, that allow for two memory accesses within one instruction cycle. Two independent memory banks can be accessed using two independent busses. One memory bank would hold either program instructions (or program and data) while the other memory bank would hold data only. With separate busses for program, data, and direct memory access (DMA), the TMS320C31 can perform concurrent program fetches, data read and write, and DMA operations. Since data and instructions reside in separate memory spaces, concurrent memory accesses are possible. The C31 architecture allows for four levels of pipelining; i.e., while an instruction is being executed, three subsequent instructions are being read, decoded, and fetched.

Operations such as addition/subtraction and multiplication are the key operations in a digital signal processor. A very important operation is the multiply/accumulate, which is useful for a number of applications requiring filtering, correlation, and spectrum analysis. Since the multiplication operation is so commonly executed and is so essential for most digital signal processing algorithms, it is to be executed in a single cycle. A typical digital signal processor contains an internal multiplier/accumulator for fast and efficient operations.

Figure 2.1 shows the functional block diagram of the TMS320C31. The TMS320C31 includes a number of registers, two blocks of internal memory, 32-bit data busses, one serial port, etc.

CPU Registers

The TMS320C31 contains the following registers, which we will use later:

1. R0–R7, eight 40-bit registers that allow for extended-precision results. These registers can store 32-bit integer and 40-bit floating-point numbers

2. AR0–AR7, eight general-purpose auxiliary registers that are commonly used for indirect memory addressing

3. IR0 and IR1, for indexing an address

4. ST, for the status of the CPU

5. SP, the system stack pointer that contains the address of the top of the stack

6. BK, to specify the block size of a circular buffer

7. IE, IF, and IOF, for interrupt enable, interrupt flag, and I/O flag, respectively

8. RC, the repeat count to specify the number of times a block of code is to be executed

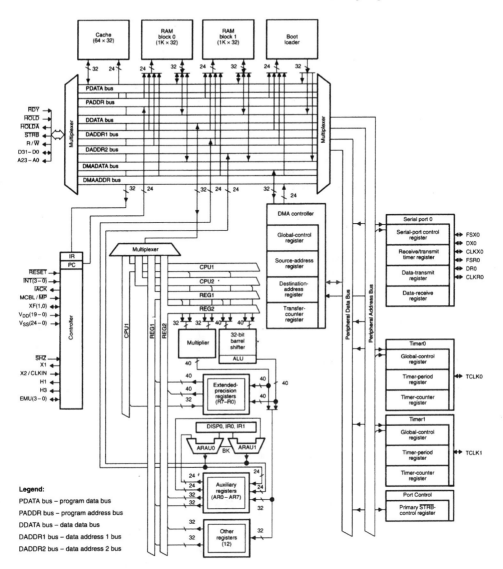

FIGURE 2.1 TMS320C31 functional block diagram (reprinted by permission of Texas Instruments).

9. RS and RE, contain the starting and ending addresses, respectively, of a block of code to be executed

10. PC, the program counter that contains the address of the next instruction to be fetched

11. DP, specifies one of 256 data pages, each page with 64K words.

The CPU registers are described in Appendix A. Several examples illustrate the utilization of these registers. For example, an extended-precision register R0 can store the 40-bit result of a multiplication of two 32-bit numbers.

Figure 2.2 shows the memory organization of the TMS320C31. RAM block 0 and RAM block 1 each contains 1K words (32-bit) of on-chip memory. However, the last 256 internal memory locations of the C31 on the DSK board are

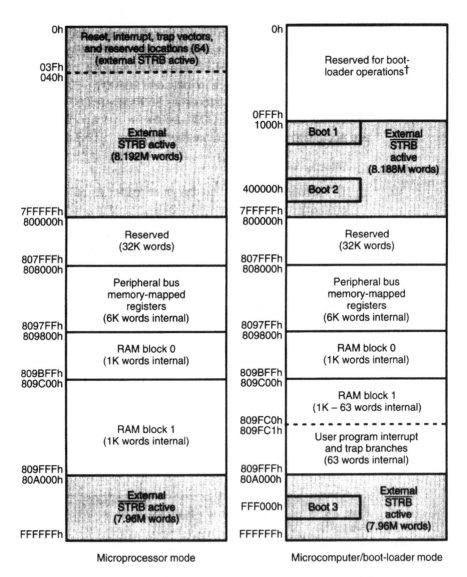

FIGURE 2.2 TMS320C31 memory organization (reprinted by permission of Texas Instruments).

used for the communications kernel and vectors. The starting address of internal memory RAM block 0 is 809800 in hex, which is half the TMS320C31 total addressable memory space of 2^{24} or 16 million 32-bit words. Figure 2.1 (top-left) shows A23-A0, which represents 24 bits of address lines. Appendix A contains the instruction set and information on registers and timers associated with the C31.

2.3 ADDRESSING MODES

Addressing modes determine how one accesses memory. They specify how data is accessed, such as retrieving an operand directly from a register or indirectly from a memory location. Several modes of addressing are available with the TMS320C31; the most commonly used mode is the indirect addressing of memory.

Indirect Addressing

Indirect memory addressing with displacement and indexing includes bit-reversed and circular modes of addressing. Registers ARn, n = 0, 1, . . . , 7 represent the eight general-purpose auxiliary registers AR0–AR7 commonly used to specify or point to memory addresses. As such, these registers are pointers. Several modes of indirect addressing follow.

a) *ARn.   This indirect mode of memory addressing is represented with the * symbol. For example, with n = 0, AR0 contains (or points to) the address of a memory location where a data value is stored; i.e., the content in memory with the address specified or pointed by AR0.

b) *ARn++(d).   The content in memory with ARn specifying the memory address. After the value in that memory location is fetched, ARn is postincremented (modified), such that the new address is the current address offset by d, or ARn+d. ARn would contain the next-higher memory address if the displacement d = 1 (d is an 8-bit unsigned integer). The index registers IR0 and IR1 are frequently utilized as the displacement d. A double minus (– –), instead of double plus, would update or postdecrement ARn to ARn-d.

c) *++ARn(d).   The content in memory with an address preincremented (modified) to ARn+d. A double minus would predecrement the memory address to ARn-d.

d) *+ARn(d).   The content in memory with the address ARn+d. ARn is not updated or modified as in the previous case.

e) *ARn++(d)%.   This is the same as in b) except that the modulus operator % (modulo arithmetic) represents a circular mode of addressing. The processor's address generation unit automatically creates the desired circular buffer, transparent to the programmer. It is used to specify an address within a circular buffer. After ARn reaches the bottom or higher address of a circular buffer, it will then point to the top address of that circular buffer when incremented next. Circular buffers are utilized extensively to implement equations that model delays in filtering and correlation, and for bit-reversal in a fast Fourier transform (FFT) algorithm. A double minus (– –) would update the address to ARn - d. If ARn is at the top address of a circular buffer, it would specify or point to the address at the bottom of the circular buffer when it is decremented next. Note that we visualize the "bottom" location of a buffer as having a higher memory address. For example, as we increment an auxiliary register or pointer to the next-higher memory address, that register will point to the subsequent *lower* memory location.

   f) *ARn++(IR0)B.   The index register or displacement d represents an offset address. This mode is similar to the previous one except that the B designates a bit-reversal process. This bit-reversal process with a reverse carry allows the necessary resequencing of data in an FFT algorithm, as illustrated in Chapter 6. ARn is updated to ARn+IR0 with reverse-carry.

   Other addressing modes [6–8] such as direct addressing are also available. For example,

```
ADDI @0x809802,R0
```

adds the data value in memory address 809802 to the value in register R0, with the result stored in R0. The symbol @ represents direct addressing.

   Another mode of addressing is register addressing. For example,

```
FIX R0,R1
```

converts a floating-point value in R0 to an equivalent integer value in R1. This instruction is very useful before sending resulting data to a DAC for output.

2.4 TMS320C3x INSTRUCTION SET

Several code segments are presented in order to become familiar with the TMS320C3x instruction set. The third-generation TMS320C3x processor has an architecture and instruction set quite different from the C1x, C2x, and C5x fixed-point processors. Even though the TMS320C3x contains a richer and more powerful set of instructions compared to these fixed-point processors, it is not any harder to program. Appendix A contains a summary of the C3x instruction set [8]. A general instruction syntax format follows:

   label   Instruction or Assembler Directive   Operand   Comment

For example, the following line of code,

```
LOOP  SUBI 1,R0  ;subtract 1 from R0
```

consists of a label (LOOP), which must start in the first column and is case-sensitive, followed by the subtract integer instruction SUBI, the operand 1,R0, and a comment. One or more blank spaces must separate each of the fields. Comments are optional and must begin with a semicolon after an operand (an instruction or an assembler directive). Comments can also start in column 1 with either a semicolon or a *. It is very instructive to read the comments in the programs discussed in this book.

Types of Instructions

1. Math Instructions to Add, Subtract, or Multiply. The instruction

```
ADDF3 R0,R2,R1
```

adds the floating-point values in registers R0 and R2 and stores the resulting floating-point value in R1. Replacing the instruction ADDF3 by SUBF3 would subtract R0 from R2, with the result stored in R1. The instruction

```
MPYF3 *AR0++,*AR1++,R0
```

multiplies the content in memory (indirect addressing) with the address specified or pointed by AR0 by the content in memory whose address is specified by AR1, and stores the resulting floating-point value in R0. It is a three-operand instruction, the "F" in MPYF represents a floating-point multiplication; an "I" would represent an integer operation. After this operation, both auxiliary registers AR0 and AR1 are postincremented by one (by default) or to the next-higher memory address. Note that AR0 and AR1 contain the two addresses of the memory locations where the two data values to be multiplied are stored.

2. Load and Store Instruction. A 32-bit word can be loaded from memory into a register or stored from a register into memory. The two instructions

```
LDI   @IN_ADDR,AR1
STF   R0,*AR2++
```

loads directly (using the symbol @) the address represented by a label IN_ADDR into the auxiliary register AR1, then stores a floating-point value R0 into memory, whose address is specified by AR2. Then, AR2 is postincremented to point at the next-higher memory address (a displacement of one by default).

Note the "I" (integer) in LDI, since an address is an integer value. We can also load a floating-point value using LDF.

   3. Input and output Instructions. The two instructions

```
LDI       @IN_ADDR,AR4
FLOAT     *AR4,R1
```

loads an (input) address represented by the label IN_ADDR directly into AR4. Then, the content in memory, whose address is specified by AR4 (IN_ADDR), is stored in the extended-precision register R1 as a floating-point value. That value might have been obtained from an analog-to-digital converter ADC as an integer. The three instructions

```
LDI   @OUT_ADDR,AR5
FIX   R0,R1
STI   R1,*AR5
```

loads an (output) address represented by OUT_ADDR directly into AR5. Then the floating-point value in R0 is converted to an equivalent integer value into R1, then stored in memory, whose address is specified by AR5. The floating-point to integer conversion instruction FIX rounds down the result. For example, the value 1.5 would become 1 and −1.5 would become −2.

   4. Branch Instructions. A standard branch instruction executes in four cycles and should be avoided whenever possible. Unconditional as well conditional branch instructions are available. A delayed branch, with or without condition, is preferable, since it can effectively execute in a single cycle. The delayed branch instruction is illustrated with the following program segment:

```
BD    FILTER
FIX   R0,R1
NOP
STI   R1,*AR5
```

The unconditional branch with delay instruction BD is to branch or go to the instruction with the label FILTER, which takes place *after* the STI R1, *AR5 instruction. Note the no operation NOP instruction. The delayed branch instruction allows the subsequent three instructions to be fetched before the program counter is modified. A conditional delayed branch instruction is illustrated with the following program segment:

```
DBNZD   AR0,FILTER
ADDF    R0,R2
FIX     R2,R2
STI     R2,*AR3
```

In the instruction DBNZD, the first D stands for decrement, the second D is for delay, and the NZ represents the condition of not zero. The auxiliary register AR0 in this case serves the function of a loop counter. AR0 is decremented by 1, and branching to the label FILTER (which could be a function) takes place after the STI instruction. Branching to FILTER would continue as long as AR0 $\geq$ 0.

5. Repeat and Parallel Instructions. a) A block of instructions can be repeated a number of times using the repeat block RPTB instruction, as illustrated in the following program segment:

```
            LDI     10,RC
            RPTB    END_BLK
            CALL    FILTER
            FIX     R0,R1
END_BLK     STI     R1,*AR5
```

The starting address (address of the repeat block instruction RPTB) of the block of code to be executed is loaded into a special repeat start address register RS and the ending address specified by the label END_BLK (which must be in column one) is loaded into the special repeat end address register RE. Note that the starting and ending address registers RS and RE are not accessed directly by the programer. The repeat counter register RC must be loaded first with the number of times the block of code is to be repeated. The block of code starting with the CALL FILTER instruction, including the store integer STI instruction, is executed 11 times (repeated RC = 10 times). Within this block of code, a subroutine FILTER is called 11 times. Execution returns each time from the FILTER subroutine to the subsequent instruction FIX R0,R1 to convert R0 from a floating-point value to an equivalent integer value R1, then stored in a memory location, whose address is specified by AR5.

b) The RPTS instruction is used to repeat the execution of a subsequent instruction a number of times, as illustrated in the following program segment:

```
            LDI     10,AR2
            RPTS    AR2
            MPYF    *AR0++,*AR1++,R0
||          ADDF3   R0,R2,R2
            ADDF    R0,R2
```

The subsequent instruction to the RPTS instruction is MPYF3, which is executed 11 times (repeated 10 times). The parallel symbol ||, which must start in column one, designates that the first addition instruction ADDF3 is in parallel with the multiply instruction; hence, it is also executed 11 times (in parallel). The second addition instruction ADDF R0,R2 is executed only once. The second R2 is not necessary in the ADDF instruction, since R2 contains the sum of R0 and R2. Note that AR2 could have been set to 10 as the operand of the repeat instruction RPTS.

The value contained in memory whose address is specified by AR0 is multi-plied by the content in memory whose address is specified by AR1, and the re-sult is stored in R0. At the same time (in parallel), R0 is added to R2 and the re-sult stored in R2. The first R0 value in the ADDF3 instruction is *not* the first resulting product, since the ADDF3 and the MPYF3 instructions are performed in parallel. The second time that the instruction ADDF3 is executed, R0 contains the resulting product of the first multiplication. The third time that ADDF3 is executed, R0 contains the resulting product of the second multiplication, and so on. The second addition instruction ADDF R0,R2 accumulates the resulting product of the last or eleventh multiplication, and is executed only once. A sec-ond R2 in that instruction is implied and can be omitted. After each multiply ex-ecution, both AR0 and AR1 are postincremented to point at the next-higher memory addresses.

The RPTS instruction is not interruptable, and if an interrupt (discussed in Chapter 3) is allowed to occur within a loop controlled by a repeat command, then RPTS must be replaced by the block repeat RPTB instruction.

6. Instructions Using Circular Buffering. A circular buffer can be utilized to model the delays in a convolution or correlation equation, and for resequenc-ing data in an FFT algorithm using bit reversal. Consider the following program segment:

```
LENGTH      .set     32
            LDI      LENGTH,BK
            RPTS     LENGTH-1
            MPYF3    *AR0++,*AR1++%,R0
||          ADDF3    R0,R2,R2
            ADDF     R0,R2
```

We will see in the next section how a directive such as .set defines the value for LENGTH as 32. The special register BK specifies the size of a circular buffer with 32 memory locations. After each multiplication, AR1 is postincre-mented to the next-higher memory location until it reaches the bottom memory address of the circular buffer. When it is next postincremented, AR1 points "back" to the initial or top (lower) memory address of the circular buffer.

Other types of instructions are available, such as *logical instructions* AND, OR, NOT, and XOR for bit manipulation, which can be useful in a decision-mak-ing process. A particular bit can be tested and a decision made based on the re-sult. A specific bit can be tested in conjunction with a shift instruction.

2.5 ASSEMBLER DIRECTIVES

Assembler directives such as .set begin with a period. An assembler directive is a message for the assembler and is not an instruction. It is resolved during the assembling process and does not occupy memory space as an instruction does. For example, the starting addresses of different sections can be specified with assembler directives, thereby eliminating the need for a linker. Consider the following program segment:

```
            .include  "prog1.asm"
            .start    ".text",0x809900
            .start    ".data",0x809C00
LENGTH      .set      32
```

A source file `prog1.asm` is "included." Several source files can be appended with the assembler directive `.include` as in C programming. The text and the data (names are case-sensitive) sections start in memory locations 0x809900 and 0x809C00, respectively. These are typical functions of a linker. LENGTH is set to 32. The following are some commonly used assembler directives and many will be illustrated through several programming examples in Section 2.7 [13]:

	.include	"prog.asm"	To include the source file `prog.asm`
A	.set	5	A is set to the value 5
B	.word	k	B is initialized to the 32-bit integer value k
C	.float	k	C is initialized to the 32-bit floating-point value k
	.text		To assemble into program memory section, equivalent to `.sect ".text"`
	.data		To assemble into data memory section, equivalent to `.sect ".data"`
	.start	"sect",addr	To start assembling at address addr. Serves the function of a linker, where sect could be `.text`
	.sect	"mysect"	To assemble into user's defined section mysect. Must have a `.start` directive before defining a section
	.entry	addr	Starting address when loading a file
	.brstart	"sect",n	Align named section (sect) as a circular buffer to the next n address boundary, with n a power of 2
	.align	K	Align section program counter (SPC) on a boundary with K being a power of 2
	.loop	n	Loop n times through a block of code
	.endloop		End of loop
	.end		End of program
	.if	cond	Assemble code if cond is not zero (true)
	.else		Otherwise (else), assemble if cond is zero (false)

.endif		End of conditional assembly of code
A .space	n	Reserve n words in current section with A as the beginning address of the reserved space
.ieee	k	k is converted to IEEE single-precision 32-bit format
.fill	45,0	To fill 45 memory locations with zero

2.6 OTHER CONSIDERATIONS

In programming the C31, a number of considerations, such as memory accesses, should be taken into account.

Conflicts

A basic instruction has four levels of pipelining: fetch, decode, read, and execute. While an instruction is being executed, the subsequent three instructions are being read, decoded, and fetched, respectively. Various stages for executing an instruction overlap and are performed in parallel. Pipelining is the overlapping of the fetch, decode, read, and execute phases of an instruction. A pipeline conflict occurs when the processing sequence of an instruction is ready to go from one pipeline level onto the next one, and that level is not yet ready to accept the transition. Fortunately, such conflicts are transparent to the programmer, and one need not to worry about that unless speed becomes a very crucial consideration [8].

Branch conflicts
Nondelayed branch instructions such as CALL, RPTB, RETS, DB cause pipelining conflicts. Since the pipeline can only handle the execution of one of these instructions, the pipeline is flushed, discarding a subsequent fetch. This flushing process prevents partial execution of a subsequent instruction. For example, a nondelayed RPTB instruction flushes the pipeline in order to load the registers RS, RE, and RC, which contain the starting address, the ending address, and the count number, respectively. With a delayed branch, execution delay can be avoided.

Register Conflicts
These conflicts occur during a read from or write to a register, within a specific group of registers (such as auxiliary registers AR0–AR7) for addressing when a register within that same group is not ready to be used. More specifically, if an instruction writes to an auxiliary register, no other auxiliary register can be decoded until the write (execution) cycle is completed. For example, a load to a register instruction followed by an instruction using that same register, i.e.,

```
LDI    K,AR0
MPYF   *AR0,R0
```

The decode phase of the MPYF instruction is delayed two cycles, since it needs the result of the preceding write to AR0. In the following example,

```
ADDI3   AR0,AR2,R1
MPYF    *AR2,R0
```

the decode stage of the MPYF instruction is delayed one cycle until AR2 is read.

Memory Conflicts

These conflicts occur because internal memory (RAM0 or RAM1) can support only two accesses per cycle. For example, two data accesses to an internal RAM block *and* a program fetch from the same internal RAM block. The C31 provides one external interface that supports only one access per cycle. Conflicts also occur when three CPU data accesses in one cycle are required. For example, a store (write) followed by two loads (reads) in parallel. The write must be completed before the two reads can be completed, delaying the reads by one cycle. The same type of conflict occurs with two writes (two stores in parallel) followed by a read.

Efficiency of Memory Access

If it is desired to have a program fetch and either one or two data accesses in one cycle, a number of alternatives can yield maximum performance within a single cycle. For example: one program access from the primary bus and two data accesses from internal RAM.

Cache

The cache is a small memory section used to store program instructions. If an instruction is being fetched from external memory, the cache feature automatically determines whether the instruction is already contained in the 64 × 32 cache memory (see Figure 2.1). If so, a "cache hit" occurs and the requested instruction is read from cache. If not, a "cache miss" occurs and the requested instruction is copied into the cache.

Since on the DSK board all program instructions are stored in internal RAM, the cache is not used. However, Appendix C describes a daughter board with 32K words each of external and flash memory that can be connected to the DSK board.

DMA

Data transfer can occur without the processor's CPU involvement. It can occur in parallel with program execution. Separate busses for program, data, and DMA allow for parallel program fetch, data read and write, and a DMA operation. For example, the C31 can perform an external program fetch, access two data values within one block of internal RAM, and use the DMA to load data to the other block of internal RAM; all within a single cycle. By performing input and output operations, the DMA can reduce the pipelining effects associated with the CPU.

Wait States

With slower peripherals such as external memory, wait states can be inserted by the programmer to accomodate access to such memory. Different numbers of wait states can be programmed and applied to different banks of memories with different speeds. As a result, slower and less-expensive memory devices can still be used.

ROM

ROM can be programmed using a PROM to store a specific application program. On-chip as well as external additional (if needed) ROM can be used for the application program as well as a boot-loader program. The TMS320C30 has an on-chip ROM while the C31 does not have one. Appendix C describes a board with external memory and flash memory connected to the DSK, and illustrates how a specific application program can be stored on flash and run without the DSK being connected to a PC host.

2.7 PROGRAMMING EXAMPLES USING TMS320C3x AND C CODE

Six programming examples are included in this chapter, using both C and TMS320C3x assembly code as well as mixed-mode with an assembly function that is called from C. Although C is more portable and more maintainable than assembly code, a C-code program does not achieve the efficiency and processing speed of a program coded in assembly. Many applications are computationally intensive and may necessitate a time-critical function to be written in assembly code. These examples will provide more familiarity with the TMS320C3x instructions, the assembler directives, and associated tools.

Example 2.1 Addition of Four Values Using TMS320C3x Code

Figure 2.3 shows the program listing ADD4.ASM for adding the four values 2, 3, 4, and 5. The assembler directive .text specifies that text or code section starts at memory location 809C00 (hex implied), which corresponds to the starting address of internal memory RAM block 1, as shown in Figure 2.2. The .float assembler directive (there is also a FLOAT instruction) defines the four values 2, 3, 4, and 5 as 32-bit floating-point constants and stored in consec-

```
;ADD4.ASM - ADD 4 FLOATING-POINT VALUES
          .start  ".text",0x809C00  ;where text begins
          .text                     ;text section
VAL_ADDR  .word   VALUES            ;starting address for values
VALUES    .float  2,3,4,5           ;the 4 values to be added
          .entry  BEGIN             ;start of code
BEGIN     LDP     VAL_ADDR          ;init to data page 128
          LDI     @VAL_ADDR,AR0     ;AR0=starting address of values
          LDF     0,R0              ;set R0=0
          RPTS    3                 ;execute next instr. 4 times
          ADDF    *AR0++,R0         ;accumulate in R0
          BR      $                 ;branch to current addr(itself)
```

FIGURE 2.3 Addition program with four values (ADD4.ASM).

utive memory location starting at the address specified by VAL_ADDR. The .entry BEGIN directive designates that the starting address of code is at BEGIN.

While it is not necessary to load or initialize the data page register to 128 using the DSK debugger, it is necessary to do so if you use another debugger such as the Code Explorer described in Appendix B. There are 256 pages (each with 64K words) of addressable memory, for a total of 2^{24} memory addresses. The instruction LDP VAL_ADDR loads the data page register DP with an address that is on page 128 (0×80 in hex). Alternatively, LDP @0x809800 would initialize the data page to 128, since 809800 (hex implied) is the starting address of internal memory (half the total memory space).

The ADDF instruction adds the content in memory starting at the address specified by AR0, which is loaded first with VAL_ADDR using the LDI instruction. The addition instruction is executed four times (repeated three times). After each execution, AR0 is postincremented to point at the next-higher memory location where the subsequent value to be added resides. The accumulation is in F0 which represents the extended-precision register R0.

This program is on the accompanying disk. Assemble it and load the resulting executable file ADD4.DSK into the debugger (after resetting the C31). Single-step through the program and verify that F0 = 14. Press F3 to display F0 in floating-point decimal. Note that F0-F7 from the CPU registers window screen represent the eight extended-precision registers R0-R7.

Example 2.2 Multiplication of Two Arrays Using TMS320C3x Code

Figure 2.4 shows the program listing MULT4.ASM, which multiplies two arrays, each containing four values.

```
;MULT4.ASM - MULTIPLY TWO ARRAYS HN AND XN EACH WITH 4 VALUES
          .start   ".data",0x809900  ;starting address of data
          .start   ".text",0x809C00  ;starting address of text
          .data                      ;data section
HN        .float   1,2,3,4           ;HN values
XN        .float   2,3,4,5           ;XN values
HN_ADDR   .word    HN                ;starting address of HN array
XN_ADDR   .word    XN                ;starting address of XN array
          .entry   BEGIN             ;start of code
          .text                      ;text section
BEGIN     LDP      HN_ADDR           ;init to data page 128
          LDI      @HN_ADDR,AR0      ;AR0=starting address of HN array
          LDI      @XN_ADDR,AR1      ;AR1=starting address of XN array
          LDF      0,R0              ;init R0=0
          LDF      0,R2              ;init R2=0
          RPTS     3                 ;execute next 2 instr. 4 times
          MPYF3    *AR0++,*AR1++,R0  ;R0=(AR0)*(AR1)
||        ADDF3    R0,R2,R2          ;in parallel with accumulation=>R2
          ADDF     R0,R2            ;last multiply result added to R2
WAIT      BR       WAIT              ;wait
```

FIGURE 2.4 Multiplication of two arrays program (MULT4.ASM).

1. The four values 1, 2, 3, and 4 in the first array HN reside in memory locations starting at the address HN_ADDR which is at 809900, where data section starts. The values 2, 3, 4, and 5 follow in the XN array. This can be verified from the **MEMORY** window screen in the debugger.

2. The starting memory addresses of the HN and XN arrays are loaded directly (using the @ symbol) into the auxiliary registers AR0 and AR1, respectively. These two addresses (809900 and 809904) are designated by HN_ADDR and XN_ADDR.

3. The content in memory (in the HN array) whose address is specified by AR0 is multiplied by the content in memory (in the XN array) whose address is specified by AR1, and the resulting product is stored in R0. Since the first addition instruction is executed in parallel with the multiply instruction, the first value in R0 that is being added is *not* the product that resulted from the first multiplication operation. That first product is not yet available to be added. The multiply instruction in parallel with the first addition instruction are executed four times. The second time that the ADDF3 instruction is executed, the result of the first product in R0 is accumulated. Hence, the second addition instruction ADDF R0,R2 is executed once to accumulate the last or fourth product.

4. The branch instruction BR causes a branch to the address specified by the WAIT label, effectively causing execution to the same instruction indefinitely (waits). An alternative instruction is BR $, which branches to the current address (itself).

5. Single-step through this program. Press F3 and verify through the C31 registers window screen that F2 = 40, since

$$R2 = (1 \times 2) + (2 \times 3) + (3 \times 4) + (4 \times 5) = 40$$

Example 2.3 Background for Digital Filtering Using TMS320C3x Code

This program example builds upon the previous two examples and provides the background necessary for implementing digital filters discussed in Chapter 4. Figure 2.5 shows a listing of the program FIR4.ASM for this example. The previous example discusses the multiplication of two sets of four numbers in two arrays or buffers HN and XN. In this example, there are three buffers:

a) an HN buffer starting at the address specified by HN_ADDR, which contains the four values 1, 2, 3, and 4

b) an input buffer IN starting at address IN_ADDR, which contains the four values 10, 0, 0, and 0

c) a special circular buffer XN_BUFFER starting at the address XN_ADDR.

The way the two arrays of numbers are being multiplied is very important, since the same method is used when implementing a digital filter called FIR, discussed in Chapter 4. In fact, this example can be readily extended to program an FIR filter. The input values in the buffer IN are transferred into the circular buffer in the same fashion as used to implement an FIR filter. This example illustrates how it is done, and in Chapter 4 it is explained why. Single-step through the program and verify the following:

1. The length of each array or buffer is four. XB_ADDR is the address at the bottom (higher-memory location) of the circular buffer XN_BUFFER, since it is specified as XN+4 - 1.

2. The .brstart assembler directive designates a circular buffer "XN_BUFFER" to be "aligned" on a 16-word boundary. The actual size of the circular buffer is four. A circular buffer must be aligned on an n-word boundary, where n is a power of two and is greater than the size of the circular buffer. If there were 65 values, the buffer would have to be aligned on a 2^7 or 128-word boundary for circular buffering. A 128-word boundary size would be required, since 128 represents the smallest power of two that is greater than 65. While a buffer could be "naturally" aligned (by luck) for circular buffer-

```
;FIR4.ASM - BACKGROUND FOR FILTER PROGRAM
            .start    ".data",0x809900  ;starting address of data
            .start    ".text",0x809C00  ;starting address of text
            .data                       ;data section
HN          .float    1,2,3,4           ;HN values
IN          .float    10,0,0,0          ;4 input values
HN_ADDR     .word     HN                ;starting address of HN array
IN_ADDR     .word     IN                ;starting address of IN array
XN_ADDR     .word     XN                ;starting address of XN buffer
XB_ADDR     .word     XN+LENGTH-1       ;last (bottom) address of XN
OUT_ADDR    .word     0x809802          ;address of output result
LENGTH      .set      4                 ;size of circular buffer
            .brstart  "XN_BUFFER",16    ;align a buffer of size 16
XN          .sect     "XN_BUFFER"       ;buffer section of XN
            .loop     LENGTH            ;loop length times
            .float    0                 ;init all XN values to zero
            .endloop                    ;end of loop
;                     +---------+       +---------+
;lower address ->     | H3 = 1  |       | X3 = 0  | <-top of XN_BUFFER
;                     +---------+       +---------+
;                     | H2 = 2  |       | X2 = 0  |
;                     +---------+       +---------+
;                     | H1 = 3  |       | X1 = 0  |
;                     +---------+       +---------+
;higher address->     | H0 = 4  |       | X0 = 10 | <-bottom of XN_BUFFER
;                     +---------+       +---------+
            .text                       ;text section
            .entry    BEGIN             ;start of code
BEGIN       LDP       HN_ADDR           ;init to data page 128
            LDI       @IN_ADDR,AR5      ;AR5=starting address of input
            LDI       @XB_ADDR,AR1      ;AR1=bottom address XN buffer
            LDI       @OUT_ADDR,AR2     ;AR2=address of result (output)
            LDI       LENGTH,BK         ;BK=4, size of circular buffer
            LDI       LENGTH,R4         ;R4=4, used as loop counter
LOOP        LDF       *AR5++,R3         ;R3=1st input value
            STF       R3,*AR1++%        ;store value at bottom of XN buffer
            LDI       @HN_ADDR,AR0      ;AR0=starting address of HN array
            CALL      FILTER            ;go to subroutine FILTER
            FIX       R2,R0             ;convert from float to integer
```

(continued on next page)

FIGURE 2.5 Background program for digital filtering (FIR4.ASM).

```
          STI     R0,*AR2++           ;store result to "output" address
          SUBI    1,R4                ;decrement loop counter R4
          BNZ     LOOP                ;branch back until R4=0
WAIT      BR      WAIT                ;wait indefinitely
;SUBROUTINE FILTER
FILTER    LDF     0,R0                ;init R0=0
          LDF     0,R2                ;init R2=0
          RPTS    LENGTH-1            ;execute next 2 instr 4 times
          MPYF3   *AR0++,*AR1++%,R0   ;R0=(AR0)*(AR1)
||        ADDF3   R0,R2,R2            ;in parallel with accumulation=>R2
          ADDF    R0,R2               ;last accumulation => R2
          RETS                        ;return from subroutine
          .end                        ;end
```

FIGURE 2.5 *(continued)*

ing, one needs to guarantee such condition. A naturally aligned buffer starts at an address in memory with the least significant four bits being zero. Otherwise, erroneous results will be produced when using such buffer for circular buffering.

3. The .loop and .endloop assembler directives specify a loop to be executed four times, and the directive .float 0 sets to zero all the memory locations within the XN_BUFFER. Such initialization method is effective, since the buffer can be initialized to zero without using instructions that occupy memory space and contribute to the program-execution time. Note that all the directives are resolved during the assembling (not the execution) process.

4. Memory locations 809900–809903 contain the four floating-point values 00000000, 01000000, 01400000, and 02000000, which are equivalent to the HN array values 1, 2, 3, and 4 [6,8]. The four values 10, 0, 0, and 0 for IN are displayed in floating-point format (03200000, 80000000, 80000000, 80000000) in memory locations 809904–809907. Note that the decimal values 1 and 0 correspond to the floating-point values 00000000 and 80000000, respectively. It is not necessary to worry about the floating-point format.

5. AR5 is loaded with 809904 (hex is implied), the starting address of the input buffer; AR1 is loaded with 809913, the bottom or higher-memory address within the circular buffer; and AR2 is loaded with 809802, the starting address for the resulting output. BK is loaded with 4, the actual size of the circular buffer (aligned within a 16-word boundary), and the value 4 is loaded into R4, which is used as a loop counter.

6. The block of code between the instruction with the label LOOP and the conditional-branch (if not zero) instruction BNZ LOOP is executed four times. Each time that this block of code is executed, the subroutine FILTER is called with the instruction CALL FILTER. Within this block of code, R3 is loaded with the content in memory (the first input value 10), whose address is specified by AR5 as 809904. The value 10 in R3 is then stored in memory location 809913, specified by AR1, which is the address at the bottom of the circular buffer (the starting or top address is at 809910). AR1 is then postincremented to point "back" to the top or lower-memory address 809910 of the circular buffer. AR0 is loaded with 809900, the starting address of the HN buffer.

7. The code within the FILTER subroutine was previously discussed in a program segment in Section 2.4. It multiplies the content in memory pointed by AR0 (the first value 1 in HN) by the content in memory (initialized before to 0) pointed by AR1 and stores the result in R0. The first resulting product is not yet available in R0 when the parallel instruction ADDF3 R0,R2,R2 is executed the first time. The multiplication instruction MPYF3 in parallel with the ADDF3 instruction are executed four times, while the second addition ADDF R0,R2 instruction is executed only once to accumulate the last product. After each multiply operation both AR0 and AR1 are postincremented to point at the next-higher memory location. After the last multiplication, AR1 increments and points back to the top address of the circular buffer. The result in R2 from the FILTER subroutine is

$$R2 = H3 \times X3 + H2 \times X2 + H1 \times X1 + H0 \times X0 = 1(0) + 2(0) + 3(0) + 4(10) = 40$$

Press F3 and verify that F2 = 40.

8. Execution is then returned to the subsequent instruction FIX R2,R0 to the CALL instruction, which converts R2 from floating-point format to integer format in R0. The result in R0 is then stored in memory, whose address is specified by AR2 as 809802. AR2 is postincremented to point at 809803 (where the second resulting value will be stored). The loop counter R4 is then decremented and execution returns to the top of the block of code within the loop with the conditional branch if not zero instruction BNZ LOOP. The program flow is then back to the address specified by the label LOOP. As you single-step the first time through the block of code within the loop, observe that the floating-point input value of 03200000, equivalent to decimal 10, is stored in memory location 809913, which is at the bottom of the circular buffer. Note that the program can be readily reloaded (after resetting) and single-step through again.

9. The block of code within the loop is now executed a second time. AR1 points to 809910, the top address of the circular buffer. AR5 points to the memory address 809905 (having been postincremented before), which con-

tains the second input value zero in the buffer IN. This value is loaded into R3, then stored in memory location 809910 (the top memory location of the circular buffer). AR1 is then postincremented to point at the memory location 809911, the second memory location of the circular buffer (from the top), which already contains the initial value of zero (initialized with the .float 0 directive). AR0 is reinitialized to the top of the HN address. The subroutine FILTER is called a second time to yield a second value for R2, or

$$R2 = H3 \times X2 + H2 \times X1 + H1 \times X0 + H0 \times X3 = 1(0) + 2(0) + 3(10) + 4(0) = 30$$

Verify that F2 = 30. After processing the FILTER subroutine a third time, the third value of R2 returned by the FILTER subroutine is

$$R2 = H3 \times X1 + H2 \times X0 + H1 \times X3 + H0 \times X2 = 1(0) + 2(10) + 3(0) + 4(0) = 20$$

and the fourth or last value of R2 is

$$R2 = H3 \times X0 + H2 \times X3 + H1 \times X2 + H0 \times X1 = 1(10) + 2(0) + 3(0) + 4(0) = 10$$

10. Note that the last multiplication operation each time that the FILTER subroutine is called involves the last HN value and the "newest" or the most recently transferred input value from the buffer IN. The first time, the newest input value of 10 was stored as X0 in the bottom memory of the circular buffer. The second time, the newest input value of zero was stored as X3 in the top memory address of the circular buffer, then as X2, and then as X1.

11. Type the command memd 0x809802 (or mem1) to display the contents in memory, starting at the address 809802, and verify the four values 40, 30, 20, and 10 in memory locations 809802–809805. Each of these values was displayed in F2. While debugger commands are not case-sensitive, the hex notation 0x is necessary within a debugger command.

This program can be modified to implement the convolution equation in Chapter 4, which represents a digital filter (see Experiment 2).

For a real-time filter implementation, each input value is obtained from an analog-to-digital converter ADC, in lieu of the input buffer IN, and stored in memory within a circular buffer in a similar fashion as in this example. The output in R2, converted from floating-point to integer, would be sent to a DAC. The block of code within the loop would be continuous, since each time that the FILTER subroutine is processed, an output value is obtained for a specific time n, where n = 0, 1, 2, 3 . . . . In Chapter 4, we will make this program more efficient. For example, a call or a branch without delay instruction takes four cycles to execute, and also it is not efficient to decrement a loop counter using the subtract instruction SUBI 1,R4.

Example 2.4 Matrix/Vector Multiplication Using TMS320C3x Code

Consider again the matrix/vector multiplication in Example 1.1 and the program listing in Figure 1.1. Even though this program looks more difficult than its C-coded counterpart in Example 1.3, it executes faster. The execution speed can be observed from _DT within the CPU register window screen. _DT displays the instruction cycle time of each instruction as you single-step through each one. Since this program executes in 100 cycles (without the last BR instruction) at 40 ns per cycle, the time is 4 μs. Note the following:

1. The (3 × 3) matrix values are in the array A, and the (3 × 1) vector values are in the array B, both in floating-point. The starting addresses of the A matrix and the B vector are specified by A_ADDR and B_ADDR, respectively. AR0, AR1, and AR2 are loaded with the starting addresses of the A matrix, the B vector, and the resulting output, respectively (809c00, 809c09, and 809c0e).

2. R4 is used as a loop counter for the outer loop between LOOPI and the instruction BNZ LOOPI, which is executed three times for each row of the matrix A. An inner loop is between LOOPJ and the instruction DB AR4,LOOPJ and is executed three times for each row in the vector B. The inner loop process continues until AR4 is less than zero, with AR4 being decremented each time, using the decrement and branch instruction DB.

3. The result is a (3 × 1) vector and each value is accumulated in R0. R0 is converted from floating-point to integer with the FIX instruction, and stored (using the STI instruction) in the memory address pointed or specified by AR2. Since AR2 is postincremented after each result is stored in memory, the three resulting values—14, 32, and 50—are stored in consecutive memory locations. Each resulting value can be verified from F0.

4. Within the outer loop, after each resulting value is obtained, the starting address of the vector B is reloaded into AR1 in preparation for the multiplication of the values in the next row of the matrix A with the column values in the vector B.

5. Type memd 0x809c0e and verify the resulting values 14, 32, and 50 stored in memory locations 809c0e-809c10.

This program can be extended to a (3 × 3) matrix A multiplying a (3 × 3) matrix B using three nested loops.

Example 2.5 Addition Using C and C-Called TMS320C3x Assembly Function

This example illustrates a main C program ADDM.C, listed in Figure 2.6, that calls an assembly function ADDMFUNC.ASM, listed in Figure 2.7. It is instruc-

```
/*ADDM.C - PROGRAM IN C CALLING A FUNCTION IN ASSEMBLY*/
extern int addmfunc();    /*external assembly function*/
int temp = 10;            /*global C variable          */
main()
{
 volatile int *IO_OUT=(volatile int *) 0x809802; /*addr for result*/
 int count;
 for (count = 0; count < 5; ++count)
   {
    *IO_OUT++=addmfunc(count); /*calls assembly function five times*/
   }
}
```

FIGURE 2.6 Addition program in C that calls an assembly function (ADDM.C).

tive to reexamine Example 1.3, a C program that multiplies a (3×3) matrix A by a (3×1) vector B. The executable file ADDM.OUT is on the accompanying disk and can be used to test this example. However, if those programs are modified, the floating-point DSP tools are needed in order to create a new executable file (see Example 1.3). These tools, version 5.0, include a C compiler, an assembler, and a linker, which were used to create the executable COFF file ADDM.OUT (on disk). They are available from Texas Instruments or other vendors. The C program was compiled using CL30 -k addm.c to create a source file addm.asm and an object file addm.obj. The compiling and linking procedures associated with a C-source code were introduced in Chapter 1 in conjunction with Example 1.3.

```
*ADDMFUNC.ASM - ASSEMBLY FUNCTION CALLED FROM C PROGRAM
FP       .set    AR3             ;frame pointer in AR3
         .global _addmfunc       ;global ref/def
         .global _temp           ;global ref/def
_addmfunc                        ;function in assembly
         PUSH    FP              ;save FP into stack
         LDI     SP,FP           ;point to start of stack
         LDI     *-FP(2),R0      ;1st count value into R0
         ADDI    @_temp,R0       ;add global variable to R0
         POP     FP              ;restore FP
         RETS                    ;return from subroutine
```

FIGURE 2.7 C-called assembly function (ADDMFUNC.ASM).

Linking

The assembly program ADDMFUNC.ASM must be assembled first with the command asm30 ADDMFUNC.ASM to create the object file ADDM-FUNC.OBJ. The linker command file ADDM.CMD (on disk) and listed in Figure 2.8 links the two object files ADDM.OBJ and ADDMFUNC.OBJ and a run-time library support file RTS30.LIB with the command LNK30 ADDM.CMD. The commands listed in the linker command file are case-sensitive. ADDM.CMD can serve as a sample linker command file. It can be readily changed to link different object files and create a COFF executable file with the -O option. The resulting executable file ADDM.OUT can be downloaded directly into the DSK and run. The extension OUT is optional within the DSK debugger, since the DSK detects an executable COFF file as opposed to an executable DSK file.

The .text section where text or code resides is specified to be in RAM0, where RAM0 is defined as a memory section starting at the address 809802 (hex implied) with a length of 0x3FE = 1022. This represents the size of internal memory block 0 (see Figure 2.2) less the first two internal-memory locations 809800 and 809801 reserved for boot loading. The second block of internal memory starts at the address 809c00 with a length of 0x3C0 = 1024 (1K).

```
/*ADDM.CMD - LINKER COMMAND FILE               */
-c                      /*using C convention       */
-stack 0x100            /*256 words stack          */
addm.obj                /*object file              */
addmfunc.obj            /*object function file     */
-O addm.out             /*executable output file   */
-l rts30.lib            /*run-time library support*/

MEMORY
  {
  RAMS: org=0x809800, len=0x2    /*boot stack       */
  RAM0: org=0x809802, len=0x3FE /*internal block 0*/
  RAM1: org=0x809C00, len=0x3C0 /*internal block 1*/
  }
SECTIONS
  {
  .text:   {} > RAM0        /*code                 */
  .cinit:  {} > RAM0        /*initialization tables*/
  .stack:  {} > RAM1        /*system stack         */
  }
```

FIGURE 2.8 Linker command file for mixed-mode addition (ADDM.CMD).

Executing ADDM.OUT

1. Consider the assembly function ADDMFUNC.ASM. Certain registers are dedicated and any of the registers R4–R7, AR4–AR7, SP, DP, and FP (used as AR3) that are modified within the assembly function must be preserved on the stack, using PUSH or PUSHF and POP or POPF in order to save and restore them, respectively. This is illustrated in ADDFUNC.ASM listed in Figure 2.7. These registers must be saved as integers, except R6 and R7. The value returned from the assembly function must be saved in R0.

2. The frame pointer FP is set in auxiliary register AR3. All C identifiers are referenced in the assembly function with underscores, such as _addmfunc and _temp. The frame pointer, with offset, is used for passing the address of an argument from the C program to the assembly function. In this case, the only argument, count, is at *-FP(2). The old frame pointer FP is at the first location in the stack.

3. Access the debugger and download the executable COFF file ADDM.OUT into the DSK (as with a DSK file). Single-step through the program. There are several instructions associated with the C31 initialization. The auxiliary register AR0 contains the starting output address 809802. The instruction STI R0, *AR0 at the address 809813 stores each output value—10, 11, 12, 13, and 14—in consecutive memory locations starting at 809802. Type the command memd 0x809802 and verify these results. They represent the initial value of 10, set in temp within the C program, added to count each time that the assembly function is called, where count takes on the values 0, 1, 2, 3, and 4. As you single-step through the program, observe the five equivalent resulting hex values a, b, c, d, and e in F0 (representing R0).

In the next example, a C program calls an assembly function to implement the $(3 \times 3) \times (3 \times 1)$ matrix/vector multiplication in Examples 1.1, 1.3, and 2.4.

Example 2.6 Matrix/Vector Multiplication Using C and C-Called TMS320C3x Assembly Function

The C program MATRIXM.C, listed in Figure 2.9 calls the TMS320C3x assembly function MATRIXMF.ASM, listed in Figure 2.10. See also Examples 1.1, 1.3, and 2.4 as well as the previous example.

1. The addresses of the A and the B arrays, and the address of the (3×1) array for the result are passed into the assembly function using the frame pointer FP, with offsets of –2, –3, and –4, respectively. Since the dedicated registers R4 and AR4 in the assembly function are used as loop counters for LOOPI and LOOPJ, respectively, they must be saved with the PUSH instruction and later restored with the POP instruction.

2. Download and run the executable COFF file MATRIXM.OUT (on disk). Single-step through much of the initialization-code section. Verify that the instruction STI R0, *AR2++ at 809833 stores each result 14, 32, and 50 into

```
/*MATRIXM.C - MATRIX/VECTOR MULT. CALLS ASSEMBLY FUNCTION */
volatile int *IO_OUTPUT = (volatile int *) 0x809802;
extern void matrixmf (float *, float *, int *); /*function*/
main()
{
 float A[3][3] = {{1,2,3},
                  {4,5,6},
                  {7,8,9}};      /*3x3 matrix A              */

   float B[3] = {1,2,3};          /*3x1 vector B              */
 matrixmf ((float *)A, (float *)B, (int *)IO_OUTPUT);
}
```

FIGURE 2.9 Matrix/vector multiplication program in C that calls an assembly function (MATRIXM.C).

```
*MATRIXMF.ASM - ASSEMBLY FUNCTION CALLED FROM C PROGRAM
FP        .set     AR3              ;frame pointer in AR3
          .global  _matrixmf        ;global ref/def
_matrixmf                           ;function in assembly
          PUSH     FP               ;save old frame pointer
          LDI      SP,FP            ;point to start of stack
          PUSH     R4               ;R4 is a dedicated C register
          PUSH     AR4              ;AR4 is a dedicated C register
          LDI      *-FP(2),AR0      ;A array
          LDI      *-FP(3),AR1      ;B array
          LDI      *-FP(4),AR2      ;pointer to IO_OUTPUT
          LDI      3,R4             ;R4 is LOOPI counter
LOOPI     LDF      0,R0             ;initialize R0
          LDI      2,AR4            ;AR4 is LOOPJ counter
LOOPJ     MPYF     *AR0++,*AR1++,R1 ;A[I,J] * B[J] = R1
          ADDF     R1,R0            ;R0 accumulates result
          DB       AR4,LOOPJ        ;decr AR4 and branch til AR4<0
          FIX      R0               ;float to integer conversion
          STI      R0,*AR2++        ;output result to IO_OUTPUT
          LDI      *-FP(3),AR1      ;reload start addr of B array
          SUBI     1,R4             ;decrement R4
          BNZ      LOOPI            ;branch while R4 <> 0
          POP      AR4              ;restore AR4
          POP      R4               ;restore R4
          POP      FP               ;restore FP
          RETS                      ;return from subroutine
```

FIGURE 2.10 C-called assembly function for matrix/vector multiplication (MATRIXMF.ASM).

memory starting at the address 809802 (as specified by AR2). Type memd 0x809802 and verify the three resulting decimal values in memory locations 809802–809804.

2.8 EXPERIMENT 2: TMS320C3x INSTRUCTIONS AND ASSOCIATED TOOLS

1. Perform/implement the addition and multiplication Examples 1.1 and 1.2.

2. Perform/implement Example 1.3 as background for implementing a digital filter.

3. Edit the FIR4.ASM program on your hard drive and save it as FIR11.ASM with the following changes:

a) Increase the size of each buffer to allow for the following 11 input values in the buffer IN: 10,000, 0, 0, . . . , 0; and 11 values (coefficients) in the buffer HN: 0, 0.0468, 0.1009, 0.1514, 0.1872, 0.2, 0.1872, 0.1514, 0.1009, 0.0468, 0.

b) Create a circular buffer of size 11 (aligned within a 16-word boundary).

c) Create an output address starting at 0x809802 labeled OUT_ADDR

d) Run this program and show that the resulting values in hexadecimal notation stored in consecutive memory starting at 0x809802 are: 0, 1d3, 3f1, 5e9, 74f, 7d0, 74f, 5e9, 3f1, 1d3, 0; and in decimal: 0, 467, 1009, 1513, 1871, 2000, 1871, 1513, 1009, 467, 0. These values are the HN values scaled by 10,000 and are stored in the files FIR11X.DAT and FIR11L.DAT. The two commands memx   0x809802 or memd 0x809802 display the content in memory in hex or signed (decimal) format, respectively, starting at the address 809802. The command:

```
SAVE FIR11X.DAT,0x809802,11,X
```

saves in hex (X for ASCII hexadecimal, or L for ASCII long) format the content of 11 consecutive memory locations starting at the address 809802 into a file FIR11X.DAT. See the debugger help menu with F1 for other options.

e) The fast Fourier transform (FFT), discussed in Chapter 6, of the 11 resulting values would yield a plot of the characteristics of a lowpass filter with a cutoff frequency at $F_s/10$. An FFT utility can be obtained from MATLAB or within the Code Explorer debugger (Appendix B).

4. Test the program FIR11.ASM with the Code Explorer described in Appendix B. From the Code Explorer debugger window screen shown in Appendix

B, select **File**, then **Load** `FIR11.DSK`. Click on run, then `halt`. The output results are in memory locations `809802-80980c`. View **Memory** and select 32-bit signed integer to display the results.

To graph the results within the debugger environment, select **View**, **Graph**, a starting address of `0x809802`, a buffer size and display size of 11, a sampling frequency of 10,000, and Frequency Domain: FFT representation. The resulting plot shows the characteristics of a lowpass filter with a cutoff frequency of $F_s/10$ or 1 kHz. Increasing the buffer size (padding with zero) will yield a graph with finer resolution. These results will again be discussed in Chapter 4, in conjunction with FIR filtering, and we will see how the filter's characteristics can be made sharper or more selective by increasing the number of the values in HN.

5. Generalize the matrix/vector multiplication program to an $(N \times M) \times (M \times K)$ matrix multiplication. Test this program by setting N, M, and K to 3, and using the same values for the *B* matrix as the values set in the *A* matrix (see Examples 1.3, 2.4, and 2.6). Implement it using either:

a) TMS320C3x assembly code or

b) C code or

c) mixed-mode with a main program in C calling an assembly function.

REFERENCES

1. R. Chassaing and D. W. Horning, *Digital Signal Processing with the TMS320C25,* Wiley, New York, 1990.

2. D. W. Horning, "An Undergraduate Digital Signal Processing Laboratory," in *Proceedings of the 1987 ASEE Annual Conference,* June 1987.

3. R. Chassaing, "Applications in Digital Signal Processing with the TMS320 Digital Signal Processor in an Undergraduate Laboratory," in *Proceedings of the 1987 ASEE Annual Conference,* June 1987.

4. K. S. Lin ed., *Digital Signal Processing Applications with the TMS320 Family: Theory, Algorithms, and Implementations,* Prentice Hall, Englewood Cliffs, NJ, Vol. 1, 1988.

5. R. Chassaing, "A Senior Project Course in Digital Signal Processing with the TMS320," *IEEE Transaction on Education,* 32, 139–145 (1989).

6. R. Chassaing, *Digital Signal Processing with C and the TMS320C30,* Wiley, New York, 1992.

7. P. Papamichalis ed., *Digital Signal Processing Applications with the TMS320 Family: Theory, Algorithms, and Implementations,* Vol. 3, Texas Instruments, Inc., Dallas, TX, 1990.

8. *TMS320C3x User's Guide,* Texas Instruments, Inc., Dallas, TX, 1997.

9. *TMS320C30 Evaluation Module Technical Reference,* Texas Instruments, Inc., Dallas, TX, 1990.

10. *Digital Signal Processing Applications with the TMS320C30 Evaluation Module: Selected Application Notes,* Texas Instruments, Inc., Dallas, TX, 1991.

11. R. Chassaing and P. Martin, "Parallel Processing with the TMS320C40," in *Proceedings of the 1995 ASEE Annual Conference,* June 1995.

12. R. Chassaing and R. Ayers, "Digital Signal Processing with the SHARC," in *Proceedings of the 1996 ASEE Annual Conference,* June 1996.

13. *TMS320C3x DSP Starter Kit User's Guide,* Texas Instruments, Inc., Dallas, TX, 1996.

3

Input and Output with the DSK

- Input and output with the Analog Interface Circuit (AIC) chip
- Communication between the PC host and the C31 DSK
- Alternative memory using external and flash memory
- Alternative input and output with a 16-bit stereo codec
- Programming examples and experiments using C and TMS320C3x code

3.1 INTRODUCTION

Typical applications using DSP techniques require at least the basic system shown in Figure 3.1, consisting of an analog input and analog output. Along the input path is an antialiasing filter for eliminating frequencies above the Nyquist frequency, defined as one-half the sampling frequency. Otherwise, aliasing occurs, in which case a signal with a frequency higher than one-half F_s is disguised as a signal with a lower frequency. The sampling theorem tells us that the sampling frequency must be at least twice the highest frequency component f in a signal, or

$$F_s > 2f$$

Hence,

$$1/T_s > 2(1/T)$$

where T_s is the sampling period, or

$$(1/2)T > T_s$$

and

51

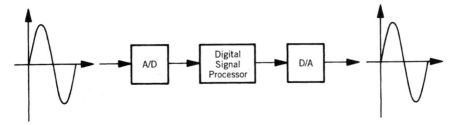

FIGURE 3.1 DSP system with input and output.

$$T_s < (1/2)T$$

The sampling period T_s must be less than one-half the period of the signal. For example, if we assume that the ear cannot detect frequencies above 20 kHz, we would sample a music signal at $F_s > 40$ kHz (typically at 44.1 kHz or 48 kHz) in order to remove frequency components higher than 20 kHz. We can then use a lowpass input filter with a bandwidth or cutoff frequency at 20 kHz to avoid aliasing.

Figure 3.2 illustrates an aliased signal. Let the sampling frequency $F_s = 4$ kHz, or a sampling period of $T_s = 0.25$ ms. It is impossible to determine whether it is the 5-kHz or the 1-kHz signal that is represented by the sequence (0, 1, 0, –1). A 5-kHz signal will appear as a 1-kHz signal; hence, the 1-kHz signal is an

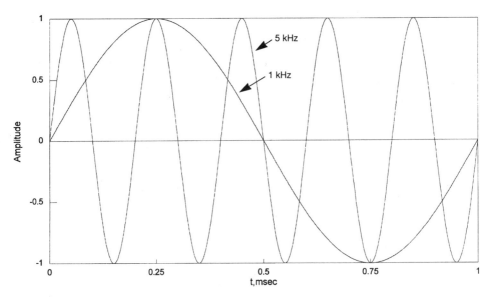

FIGURE 3.2 Aliased sinusoidal waveform.

aliased signal. Similarly, a 9-kHz signal would also appear as a 1-kHz aliased signal. We will verify this aliasing phenomenon with a programming example in Section 3.4.

The A/D converts the input analog signal to a digital representation to be processed by the digital signal processor. The maximum level of the input signal to be converted is determined by the specific analog-to-digital converter (ADC). Discrete levels or steps are used to represent the input signal. The number of steps is based on the range of the input-signal level and the number of bits of the ADC. After the captured signal is processed, the result needs to be sent to the outside world. Along the output path is a digital-to-analog converter (DAC) that performs the reverse operation of the ADC, with different output levels produced by the DAC based on its input. An output filter smooths out or reconstructs the steps into an equivalent analog signal.

3.2 THE ANALOG INTERFACE CIRCUIT (AIC) CHIP

The DSK board includes an analog interface circuit (AIC) chip that connects to the serial port on the C31. The AIC contains an ADC and a DAC as well as switched-capacitor antialiasing input filter and reconstruction output filter, all on a single C-MOS chip. Figure 3.3 shows the TLC32040 AIC functional block diagram with two inputs, one output, 14-bit ADC and DAC, and input and output filters. Programmable sampling rates with a maximum of 20 kHz for maxi-

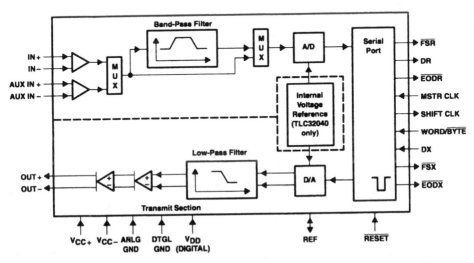

FIGURE 3.3 TLC32040 AIC functional block diagram (reprinted by permission of Texas Instruments).

mum performance are possible, although higher sampling rates for audio applications can be obtained, as described later.

The TLC32040 AIC is a member of the TLC3204x family of analog interface circuit chips [1–5]. The evaluation module (EVM) contains the TLC32044 AIC, which has an input lowpass filter as well as a bypassable highpass input filter in lieu of the bandpass input filter shown in Figure 3.3 [3].

The AIC primary input IN can be accessed from an RCA connector on the DSK board. The AIC auxiliary input AUX IN is accessed through pin 3 from the 32-pin connector JP3 along the edge of the DSK board. The input bandpass filter is bypassable and can be programmed for a desired cutoff frequency or bandwidth based on the sampling frequency. The output reconstruction lowpass filter is fixed.

AIC Control

Data transmission occurs through the data receive (DR) and the data transmit (DX) registers, two of the AIC's serial port registers. The AIC is controlled through the data transmit register. The two least significant bits (LSBs) are used for communication functions. When the two LSBs are zeros, normal transmission occurs, and when they are ones, secondary communication takes place. Secondary communication initializes and controls the AIC, allowing one secondary transmission before switching back. Figure 3.4 shows the AIC secondary communication protocol. Control functions are initiated by writing to several of the AIC's registers. Certain AIC specifications, such as input port and input filter, are obtained using the control register. For example, as shown in Figure 3.4, setting bit d2 and d4 to ones in the control register, inserts the AIC's input bandpass filter and enables the auxiliary input AUX IN, respectively.

Registers A and B on the AIC designate the location of control. The A registers consist of TA and RA and represent filter control, and the B registers consist

secondary DX serial communication protocol

x x I ← to TA register → I x x I ← to RA register → I 0 0	d13 and d6 are MSBs (unsigned binary)	
x I ← to TA' register → I x I ← to RA' register → I 0 1	d14 and d7 are 2's complement sign bits	
x I ← to TB register → I x I ← to RB register → I 1 0	d14 and d7 are MSBs (unsigned binary)	
x x x x x x x x d7 d6 d5 d4 d3 d2 1 1		
Control register	d2 = 0/1 deletes/inserts the bandpass filter	
	d3 = 0/1 disables/enables the loopback function	
	d4 = 0/1 disables/enables the AUX IN+ and AUX IN− terminals	
	d5 = 0/1 asynchronous/synchronous transmit receive sections	
	d6 = 0/1 gain control bits (see gain control section)	
	d7 = 0/1 gain control bits (see gain control section)	

FIGURE 3.4 AIC secondary communication protocol (reprinted by permission of Texas Instruments).

of TB and RB registers and represent the A/D and D/A control. These registers are associated with the AIC's internal timing configuration [1]. The bit locations for the transmit and receive registers TA and RA are:

bits 0–1   0,0
bits 2–6   RA
bits 7–8   don't care (x)
bits 9–13   TA
bits 14–15   don't care (x)

The bit locations for the transmit and receive registers TB and RB are:

bits 0–1   0,1
bits 2–7   RB
bit 8   don't care (x)
bits 9–14   TB
bit 15   don't care

The AIC can be configured for a specified sampling frequency and filter bandwidth by requesting secondary communication and loading ones in the first two LSBs. Secondary communication follows a primary communication that has the two LSBs set to ones. The following sequence of data is loaded to the serial port data transmit register and sets the two LSBs to one for each secondary communication request:

a) 0x3 (or 3h) to request secondary communication
b) value for the A registers
c) 0x3 to request secondary communication again
d) value for the B registers
e) 0x3 to request secondary communication a third time
f) value to configure the control register

We can now proceed to find the A and B values in order to achieve a desired sampling frequency and input filter bandwidth BW.

Calculating Values for A and B for a Desired F_s and Filter BW

The C31 DSK has a 50-MHz input clock (CLKIN) that can generate a maximum timer frequency of MCLK = (CLKIN/4) = 12.5 MHz, which is above the AIC's maximum master clock frequency of 10 MHz specified for maximum performance. The AIC master clock MCLK can be accessed and measured from pin 8 on JP1 [4]. To achieve maximum performance with the AIC, we can divide the input clock by 8, or

$$MCLK = CLKIN/8 = (50 \text{ MHz}/8) = 6.25 \text{ MHz} \qquad (3.1)$$

The switched-capacitor filter frequency (SCF) is related to the A transmit register, or

$$SCF = MCLK/(2 \times TA) \qquad (3.2)$$

and the sampling frequency is related to the transmit A and B registers, or

$$F_s = MCLK/(2 \times TA \times TB) \qquad (3.3)$$

The input filter bandwidth or cutoff frequency is set at 3600 Hz for an SCF of 288 kHz [1]. A new SCF will result for a different BW. The following calculations illustrate the above and how to find the A and B values to set the AIC.

1. $F_s = 8 \text{ kHz (Desired)}$

The desired cutoff frequency of the input antialiasing filter is 3600 Hz for an SCF of 288 kHz. From (3.2)

$$TA = MCLK/(2 \times SCF) = 6.25 \text{ MHz}/(2 \times 288 \text{ kHz}) = 10.85$$
$$\cong 11 = (01011)_b \qquad (3.4)$$

From (3.3)

$$TB = MCLK/(2 \times TA \times F_s) = 6.25 \text{ MHz}/(2 \times 11 \times 8,000) = 35.51$$
$$\cong 36 = (100100)_b$$

From (3.4), the actual SCF is

$$SCF = 6.25 \text{ MHz}/(2 \times TA) = 284.09 \text{ kHz}$$

The actual cutoff frequency or input filter bandwidth is shifted accordingly, or

$$BW = 3600(\text{New SCF/Set SCF})$$
$$= 3600 \ (284.09 \text{ kHz}/288 \text{ kHz}) = 3551.14 \text{ Hz}$$

The actual sampling frequency is then

$$F_s = 6.25 \text{ MHz}/(2 \times TA \times TB) = 6.25 \text{ MHz}/(2 \times 11 \times 36) = 7891.41 \text{ Hz}$$

From Figure 3.4, using the bit locations for the control register, and setting TA =
RA, with 5 bits for TA, 6 bits for TB, and x for don't care,

$$0 \ 0 \ 0 \ 1 \ 0 \ 1 \ 1 \ 0 \ 0 \ 0 \ 1 \ 0 \ 1 \ 1 \ 0 \ 0 \Rightarrow 162\text{Ch}$$
$$\text{x x} \ | \quad \text{TA} \quad | \ \text{x x} \ | \quad \text{RA} \quad |$$

Separating the bits into nibbles or groups of four, A = 162Ch. Similarly, with
TB = RB

$$0 \ 1 \ 0 \ 0 \ 1 \ 0 \ 0 \ 0 \ 1 \ 0 \ 0 \ 1 \ 0 \ 0 \ 1 \ 0 \Rightarrow 4892\text{h}$$
$$\text{x} \ | \quad \text{TB} \quad | \ \text{x} \ | \quad \text{RB} \quad |$$

B = 4892h. These values can be verified using a utility program AICCALC
included with the DSK software tools.

2. F_s = 10 kHz (Desired)

Using the same cutoff frequency or BW for the input antialiasing filter as previously obtained with F_s = 8 kHz, TA = 11. Then,

$$\text{TB} \ = 6.25 \ \text{MHz}/(2 \times 11 \times 10{,}000) = 28.41 \cong 28 = (011100)_b$$

The actual sampling frequency is

$$F_s = 6.25 \ \text{MHz}/(2 \times \text{TA} \times \text{TB}) = 6.25 \ \text{MHz}/(2 \times 11 \times 28) = 10{,}146 \ \text{Hz}$$

The B value is then

$$0 \ 0 \ 1 \ 1 \ 1 \ 0 \ 0 \ 0 \ 0 \ 1 \ 1 \ 1 \ 0 \ 0 \ 1 \ 0 \Rightarrow 3872\text{h}$$
$$\text{x} \ | \quad \text{TB} \quad | \ \text{x} \ | \quad \text{RB} \quad |$$

or B = 3872h.

3. F_s = 20 kHz (Desired)

Let BW = 8000 Hz (desired). Since the bandwidth is

$$\text{BW} = 3600(\text{New SCF/Set SCF})$$

the new switched-capacitor filter frequency is

$$\text{SCF} = 8000(288 \ \text{K})/3600 = 640 \ \text{kHz}$$

and the TA and TB register values are

$$TA = 6.25 \text{ MHz}/(2 \times 640 \text{ K}) = 4.88 \cong 5 = (00101)_b$$

$$TB = 6.25 \text{ MHz}/(2 \times 5 \times 20000) = 31.25 \cong 31 = (011111)_b$$

The actual SCF is

$$SCF = 6.25 \text{ MHz}/(2 \times 5) = 625 \text{ kHz}$$

The actual bandwidth is

$$BW = 3,600(625 \text{ K}/288 \text{ K}) = 7812.5 \text{ Hz}$$

The actual sampling frequency is

$$F_s = 6.25 \text{ MHz}/(2 \times 5 \times 31) = 20,161.29 \text{ Hz}$$

The A value is then

```
0 0 0 0 1 0 1 0 0 0 0 1 0 1 0 0  ⇒ 0A14h
x x |   TA    | x x |   RA   |
```

or A = 0A14h and the B value is

```
0 0 1 1 1 1 1 0 0 1 1 1 1 1 1 0  ⇒ 3E7Eh
x |   TB    | x |   RB    |
```

or B = 3E7Eh. The A and B registers for four different sampling rates follow:

(Desired)F_s, Hz	(Actual) F_s	A	B
8,000	7,891.41	0x162C	0x4892
10,000	10,146	0x162C	0x3872
16,000	15,943	0x0E1C	0x3872
20,000	20,161.29	0x0A14	0x3E7E

For $F_s = 16$ kHz, the actual BW = 5580 Hz.

There is an additional set of registers TA′ and RA′ that can be used for fine-tuning the sampling rate and filter bandwidth. In Section 3.4, we will set the A and B values in the program examples in order to obtain a desired F_s and BW.

3.3 INTERRUPTS AND PERIPHERALS

Interrupts

The TMS320C31 supports both internal and external interrupts that can interrupt the CPU or the DMA, as well as a nonmaskable external reset interrupt [6]. Figure A.1 in Appendix A shows the global interrupt enable (GIE) bit register, within the status register (ST), that controls all CPU interrupts. The GIE bit is set to one to enable an interrupt. To disable an interrupt, disable the interrupt enable (IE) register shown in Figure A.2 by setting it to zero, then set the GIE bit also to zero. Figure A.3 shows the memory-mapped locations used for interrupts [6].

Timers

The TMS320C31 supports two timers that can be used to count external events. They provide the timing necessary to signal an ADC to start conversion. Figure A.4 in Appendix A shows the peripheral bus memory-mapped registers. The timer global control register (Figure A.5) at memory location 808020 monitors the timer's status, and the timer period register at the memory address 808028 specifies the timer's frequency. The timer counter register at memory location 808024 contains the value of the incrementing counter. When the value of the period register equals that of the timer counter register, the counter register resets to zero. At reset, both the timer counter and the period registers are set to zero. We will use these registers to set a desired interrupt rate, effectively achieving a desired F_s.

Programming examples will further illustrate the use of an interrupt generated internally with a timer. Section 8.4 describes a project to control the amplitude of a generated sinewave using both internal and external interrupts.

Serial Port

The TMS320C31 supports one serial port (the C30 has two) with a set of control registers as shown in Figure A.4. Figure A.6 shows the serial port global control register format. The AIC on board the DSK connects to the C31 serial port, through a 22-pin connector jumper block JP1 that connects the C31 signals to the AIC. These jumpers can be removed to disconnect the on-board AIC from the C31 and use JP1 to access the C31 signals and interface to an external board. Appendix D describes a board that contains a CS4216 (or CS4218) 16-bit codec that interfaces to the C31 signals through the 22-pin connector JP1.

AIC Data Configuration

The following registers are set in order to initialize the AIC:

Register	Address	Command
Timer 0 period	0x808028	Load 0x1
Timer 0 global control	0x808020	Load 0x3C1
I/O flag	IOF	Load 0x2
SP0 transmit port control	0x808042	Load 0x131
SP0 receive port control	0x808043	Load 0x131
SP0 global control	0x808040	Load 0x0E970300
SP0 data transmit	0x808048	Load 0x0
I/O	IOF	Load 0x6
Interrupt flag	IF	Load 0x0
Interrupt enable	IE	OR 0x10
Status register	ST	OR 0x2000

In the next section, we will illustrate how to configure the AIC through programming examples.

3.4 PROGRAMMING EXAMPLES USING TMS320C3x AND C CODE

Several programming examples using both assembly and C code illustrate interrupts and I/O communications with the AIC. We developed a program in both assembly and C that contains several routines to communicate with the AIC for input and output. Such program can be used as an AIC "communication box." In Example 1.2, we generated a real-time sinusoid with the program SINE4P.ASM. The program SINE4P.ASM "includes" the AIC communication program AICCOM31.ASM that contains the AIC routines for initialization and input/output capabilities.

Example 3.1 Internal Interrupt Using TMS320C3x Code

Figure 3.5 shows a listing of the program INTERR.ASM that illustrates an interrupt generated internally by the C31 timer 0. The rate at which an interrupt occurs is determined without the use of the AIC. Consider the following from the program.

1. The interrupt rate is determined by a value set in the period register, or rate = 12.5 MHz/(2 × period)

2. The period register at the memory address 808028 and the global register at the memory address 808020 are initialized. Bit 8 within the interrupt enable (IE) register TINT0 is enabled. Appendix A contains information on these registers.

```
;INTERR.ASM - DEMONSTRATES INTERNAL INTERRUPT WITHOUT THE AIC
        .start  "intsect",0x809FC9 ;starting address for interrupt
        .start  ".text",0x809A00    ;starting address for text
        .start  ".data",0x809C00    ;starting address for data
        .sect   "intsect"       ;section for interrupt
        BR      ISR             ;interrupt vector TINT0
        .data                   ;data section
PERIOD   .word  2000H           ;interrupt rate=12.5MHz/(2*PERIOD)
IE_REG   .word  100H            ;enable timer 0 (TINT0) for interrupt
PER_ADDR .word  808028H         ;(TLCK0) period register location
TCNTL    .word  2C1H            ;control register value
ST_REG   .word  2000H           ;set status register
OUTPUT   .word  0xA             ;initial output value
OUT_ADDR .word  0x809A30        ;output address
STACKS   .word  809F00h         ;init stack pointer
        .entry  BEGIN           ;start of code
        .text                   ;assemble into text section
BEGIN    LDP    STACKS          ;init data page
         LDI    @STACKS,SP      ;SP -> 0809F00h
         LDI    @PER_ADDR,AR0   ;TINT0 period register =>AR0
         LDI    @OUT_ADDR,AR1   ;output address =>AR1
         LDI    @PERIOD,R0      ;period value => R0
         STI    R0,*AR0—(8)     ;set TLCK0 period @ 808028H
         LDI    @TCNTL,R0       ;control register value =>R0
         STI    R0,*AR0         ;set TLCK0 global control @ 808020H
         LDI    @OUTPUT,R0      ;R0 = output value
         OR     @IE_REG,IE      ;enable TINT0 interrupt bit 8
WAIT     IDLE                   ;wait for interrupt
         BR     WAIT            ;branch to WAIT until interrupt
;   INTERRUPT VECTOR
ISR      ADDI   2,R0            ;increment output value by 2
         STI    R0,*AR1++       ;store output value
         RETI                   ;return from interrupt
         .end                   ;end
```

FIGURE 3.5 Interrupt program using TMS320C3x code (INTERR.ASM).

3. Execution continues within a loop containing the two instructions WAIT IDLE and BR WAIT until an interrupt occurs. The counter register at memory address 808024 increments from 0, 1, ..., until it reaches the period value of 2000h set in the period register at which time an interrupt occurs. The counter register is reset to zero and is incremented again.

4. On interrupt, execution proceeds to the interrupt service routine (ISR). An initial value of 0xA = 10 (decimal) set as output is incremented by two (within the interrupt service routine) and the result stored in memory location 809a30, the starting output address specified by OUT_ADDR.

5. Execution returns to the WAIT loop until the next interrupt occurs.

6. Run this program for one or two seconds, then stop/halt execution. Type memd 0x809a30 to verify the output values 12, 14, 16, 18, ... The C31 should be reset first, before displaying the output values, if an old version of the DSK tools is used.

Due to the interrupt structure, it is not possible to single-step and observe the counter register at 808024 incrementing, or to observe the sequence of the program counter PC illustrating the instruction to be executed next, specifically when the timer counter register equals the period register value. A modified version of this program can be single-stepped through using a simulator available from Texas Instruments [3]. The simulator is a software program similar in function to the debugger but which models and does not require the C31. With a debugger, the executable file is downloaded into an actual C31 chip.

Example 3.2 Sine Generation with AIC Data Using TMS320C3x Code

Figure 3.6 shows a listing of the program SINEALL.ASM that generates a sinusoid with four points in a look-up table and contains the necessary code to communicate with the AIC. This example can serve as a sample program that illustrates how to integrate AIC communication data directly within a specific program.

Example 1.2 illustrates a sine generation program using a table look-up procedure with four points that calls the AIC routines included in a separate file

```
;SINEALL.ASM - GENERATES A SINE WITH 4 POINTS USING AIC POLLING
            .start    ".text",0x809900   ;starting addr for code
            .start    ".data",0x809c00   ;starting addr for data
            .data                        ;data section
PBASE       .word     808000h            ;peripheral base address
SETSP       .word     0E970300h          ;serial port set-up data
ATABLE      .word     AICSEC             ;SP0 AIC init table addr
AICSEC      .word     162Ch,1h,4892h,67h ;Fs = 8 kHz
SINE_ADDR   .word     SINE_VAL       ;address of sine values
            .brstart  "SINE_BUFF",8  ;size of sine table
```

(continued on next page)

FIGURE 3.6 Sine generation program with AIC data incorporated (SINEALL.ASM).

```
SINE_VAL  .word   0,1000,0,-1000  ;sine values
LENGTH    .set    4               ;length of circular buffer
          .entry  BEGIN           ;start of code
          .text                   ;assemble into text section
BEGIN     LDP     AICSEC          ;init to data page 128
          LDI     @PBASE,AR0      ;AR0=peripheral base address
          LDI     1h,R0           ;Timer CLK=H1/2*(AIC master CLK)
          STI     R0,*+AR0(28h)   ;timer period reg(TCLK0=6.25MHZ)
          LDI     03C1h,R0        ;to init timer global register
          STI     R0,*+AR0(20h)   ;reset timer
          LDI     62h,IOF         ;AIC reset = 0
          LDI     @ATABLE,AR1     ;AR1=AIC init data
          RPTS    99              ;repeat next instr 100 times
          NOP                     ;keep IOF low for a while
          LDI     131h,R0         ;X & R port control register data
          STI     R0,*+AR0(42h)   ;FSX/DX/CLKX=SP operational pins
          STI     R0,*+AR0(43h)   ;FSR/DR/CLKR=SP operational pins
          LDI     @SETSP,R0       ;RESET->SP:16 bits,ext clks,std mode
          STI     R0,*+AR0(40h)   ;FSX=output & INT enable SP global reg
          LDI     0,R0            ;R0=0
          STI     R0,*+AR0(48h)   ;clear serial port XMIT register
          OR      06h,IOF         ;bring AIC out of reset
          LDI     03h,RC          ;RC=3 to transmit 4 values
          RPTB    SECEND          ;repeat 4 data transmit of sec com
          CALL    TWAIT           ;wait for data transmit
          LDI     03h,R0          ;valuefor secondary XMIT request
          STI     R0,*+AR0(48h)   ;secondary XMIT request to AIC
          CALL    TWAIT           ;wait for data transmit
          LDI     *AR1++(1),R0    ;R0=next AIC data
SECEND    STI     R0,*+AR0(48h)   ;DTR=curent AIC data
          LDI     LENGTH,BK       ;BK=size of circular buffer
          LDI     @SINE_ADDR,AR1  ;AR1=address of sine values
LOOP      LDI     *AR1++%,R7      ;R7=table value
          CALL    TWAIT           ;wait for data transmit
          LSH     2,R7            ;Two LSB MUST = 0 for primary AIC com
          STI     R7,*+AR0(48h)   ;DTR=next data for AIC D/A
          BR      LOOP            ;branch back to LOOP
TWAIT     LDI     *+AR0(40h),R0   ;R0=content of SP global control reg
          AND     02h,R0          ;see if transmit buffer is ready
          BZ      TWAIT           ;if not ready, try again
          RETS                    ;branch from subroutine
```

FIGURE 3.6 *(continued)*

AICCOM31.ASM. A C version of the AIC communication program is described later. These routines enable the initialization of the AIC for input/output. While it is more efficient to integrate these AIC routines within each specific program for faster execution, it is more convenient to use these routines as a "black box," as was done in Example 1.2.

Appendix A describes a number of special registers on the C31 that are available for communicating with the AIC. Assemble and run SINEALL.ASM to verify a generated output sinusoid with a frequency of $f = F_s/4 = 2$ kHz. Consider the following from the program.

1. The values in AICSEC specify a sampling rate of 8 kHz with a bandwidth of 3551 Hz. The DAC output rate is the same as the input ADC rate (no input is used in this program example). The AIC master clock is set to 6.25 MHz with the instruction LDI 1,R0 with R0 stored in the timer-period register. Example 1.2 illustrates how the AIC master clock can be changed with that instruction. For example, a value of two in the timer-period register with LDI 2,R0 reduces the AIC master clock to 3.125 MHz, and effectively also reduces the sampling rate by two. The AIC master clock frequency can be verified from pin 8 on the DSK board connector JP1. Figure A.4 in Appendix A shows the memory-mapped timer locations. The second value of 1h in AICSEC sets the registers TA′ and RA′ on the AIC for fine-tuning the sampling frequency (though not used).

2. The following registers are initialized: the global control register at memory location 808020 (using timer 0), the IOF register, the serial port control registers at 808042 and 808043, the serial port global control register at 808040, and the data transmit register at 808048 (see Figures A4–A8).

3. By initializing the timer global control register with 0x3C1, bit 8 (C/$\overline{P}$) in Figure A.5 is set to one and the clock mode is chosen (not the pulse mode), which allows for an external output of 50% duty cycle.

4. Request for secondary communication is made through the data transmit register to transmit the four values set in AICSEC that specify a sampling rate of 8 kHz, the filter's BW, the AIC primary input IN, and the insertion of the AIC input bandpass filter.

5. The sequence of four values represents a sine waveform, set in SINE_VAL, and are then transmitted through the data transmit register at memory location 808048 (Figure A.4) through a polling procedure within the TWAIT routine.

6. The IOF register is kept low for a while. The AIC reset pin is connected to the C31 XF0 pin (see Figure A.8).

7. The serial port global control register is loaded with 0E970300 and causes the following (Figures A.4 and A.6):

    a. Configures FSX as input

    b. Disables handshake mode

    c. Sets both transmit and receive sync pulses to variable rate

    d. Sets both transmit and receive frame sync modes to standard mode

    e. Sets all clocks and data interface pin polarities to active high

    f. Sets all frame sync pulses to active low

    g. Transfers 16-bit data

    h. Disables all interrupts except the transmit interrupt

    i. Starts serial port operations

    j. Loads the data transmit register with an initial value of zero

8. Within the block of code or loop starting at the instruction RPTB SECEND and ending at the label SECEND, the first three lines of code load the data transmit register with the primary communication data for the AIC and the subsequent three lines of code load the data transmit register with the secondary communication data for the AIC.

9. The AIC will issue the transmit sync pulse to the C31 to start primary communication. After the data is received, the AIC uses bits 2–15 as D/A data and bits 0–1 as control data. Both control bits being set to 1 will cause the AIC to issue another transmit sync pulse (four AIC shift clock cycles after the primary communication ends) to the C31 to start secondary communication. After the secondary communication data is received, the AIC uses bits 0–1 to control the register that will be loaded and bits 2–15 as the data that will be loaded in the AIC register. The next data received by the AIC will be treated as primary communication data. All primary communications are performed at an interval that is determined by the A/D and D/A conversion rates.

10. The AIC is ready for transmission of a new word when bit 1 of the serial port global control register XRDY is set to 1 (Figure A.6), otherwise wait.

Example 3.3 Loop/Echo with AIC Routines in Separate File, Using TMS320C3x Code

This example illustrates input and output with the AIC and the effects of aliasing. Figure 3.7 shows a loop or echo program LOOP.ASM that "includes" the program AICCOM31.ASM shown in Figure 3.8. This separate program AICCOM31.ASM contains the AIC communication routines (see also SINEALL.ASM). This program was introduced in Example 1.2 in Chapter 1. It is instructive to read the comments in these programs. The program AIC-COM31.ASM includes options to achieve a data conversion rate using either interrupt or polling, and to access the primary and auxiliary inputs. Consider the following.

1. The routine AICSET in AICCOM31.ASM is called to initialize the AIC, followed by calling the routine AICIO_P for input and output using a polling

```
;LOOP.ASM - LOOP PROGRAM. CALLS AIC ROUTINES IN AICCOM31.ASM
        .start    ".text",0x809900   ;starting address for text
        .start    ".data",0x809C00   ;starting address for data
        .include  "AICCOM31.ASM"     ;AIC communication routines
        .data                        ;data section
AICSEC  .word     162Ch,1h,4892h,67h ;Fs = 8 kHz
        .text                        ;text section
        .entry    BEGIN              ;start of code
BEGIN   LDP       AICSEC             ;init to data page 128
        CALL      AICSET             ;init AIC
LOOP    CALL      AICIO_P            ;R6 = input, R7 = output
        LDI       R6,R7              ;output R7=new input in R6
        BR        LOOP               ;loop continuously
        .end                         ;end
```

FIGURE 3.7 Loop/echo program using TMS320C3x code (LOOP.ASM).

```
*AICCOM31.ASM - AIC COMMUNICATION ROUTINES - POLLING OR INTERRUPT
        .data                      ;assemble into data section
PBASE   .word    808000h           ;peripheral base address
SETSP   .word    0E970300h         ;serial port set-up data
ATABLE  .word    AICSEC            ;SP0 AIC init table address
        .text                      ;assemble into text section
AICSET  PUSH     AR0               ;save AR0
        PUSH     AR1               ;save AR1
        PUSH     R0                ;save R0
        PUSH     R1                ;save R1
        LDI      @PBASE,AR0        ;AR0 -> 808000h
        LDI      1,R0              ;timer CLK=H1/2*(AIC master CLK)
        STI      R0,*+AR0(28h)     ;timer period reg(TCLK0=6.25 MHZ)
        LDI      03C1h,R0          ;init timer global register
        STI      R0,*+AR0(20h)     ;reset timer
        LDI      62h,IOF           ;AIC reset = 0
        LDI      @ATABLE,AR1       ;AR1 -> AIC init data
        RPTS     99                ;repeat next instr 100 times
        NOP                        ;keep IOF low for a while
        LDI      131h,R0           ;X & R port control register data
        STI      R0,*+AR0(42h)     ;FSX/DX/CLKX=SP operational pins
        STI      R0,*+AR0(43h)     ;FSR/DR/CLKR=SP operational pins
```

(continued on next page)

FIGURE 3.8 AIC communication program (AICCOM31.ASM).

```
            LDI     @SETSP,R0        ;RESET->SP:16 bits,ext clks,std mode
            STI     R0,*+AR0(40h)    ;FSX=output&INT enable SP global reg
            LDI     0,R0             ;R0 = 0
            STI     R0,*+AR0(48h)    ;clear serial port XMIT register
            OR      06h,IOF          ;bring AIC out of reset
            LDI     03h,RC           ;RC=3 to transmit 4 values
            RPTB    SECEND           ;repeat 4 data transmit of sec com
            CALL    TWAIT            ;wait for data transmit
            LDI     03h,R0           ;value for secondary XMIT request
            STI     R0,*+AR0(48h)    ;secondary XMIT request to AIC
            CALL    TWAIT            ;wait for data transmit
            LDI     *AR1++(1),R0     ;AR1 -> next AIC init data
SECEND      STI     R0,*+AR0(48h)    ;DTR = current AIC data
            POP     R1               ;restore R1
            POP     R0               ;restore R0
            POP     AR1              ;restore AR1
            POP     AR0              ;restore AR0
            RETS                     ;return from subroutine
AICSET_I                             ;--CONFIG FOR INTERRUPT ------
            CALL    AICSET           ;call AICSET routine
            LDI     0h,IF            ;clear IF register
            OR      10h,IE           ;enable EXINT0 CPU interrupt
            OR      2000h,ST         ;global interrupt enable
            RETS                     ;return from subroutine
;------------TRANSMIT WAIT ROUTINE--------------
TWAIT       PUSH    AR0              ;save AR0
            PUSH    R0               ;save R0
            LDI     @PBASE,AR0       ;AR0 -> 0808000h
TW1         LDI     *+AR0(40h),R0    ;R0=content of SP global control reg
            AND     02h,R0           ;see if transmit buffer is ready
            BZ      TW1              ;if not ready, try again
            POP     R0               ;restore R0
            POP     AR0              ;restore AR0
            RETS                     ;return from subroutine
;------------AIC TRANSFER ROUTINE------------
AICIO_I LDI     R7,R6            ;copy output to modify for AIC
            LSH     2,R6             ;two LSB must=0 for primary AIC comm
IO          PUSH    AR0              ;save AR0
            LDI     @PBASE,AR0       ;AR0 -> 0808000h
            STI     R6,*+AR0(48h)    ;DTR = next data for AIC D/A
```

(continued on next page)

FIGURE 3.8 *(continued)*

```
        LDI    *+AR0(4Ch),R6   ;R6 = DRR data from AIC A/D
        LSH    16,R6           ;left shift for sign extension
        ASH    -18,R6          ;right shift keeping sign
        POP    AR0             ;restore AR0
        RETS                   ;return from subroutine
;————————AIC POLLING ROUTINE————————--
AICIO_P CALL   TWAIT           ;wait for data to be transferred
        CALL   AICIO_I         ;call AIC transfer routine
        RETS                   ;return from subroutine
SW_IO   PUSH   AR0             ;save AR0
        LDI    @PBASE,AR0      ;AR0 -> 0808000h
        LDI    R7,R6           ;copy output to modify for AIC
        LSH    2,R6            ;prepare for secondary AIC com
        OR     03h,R6          ;set two LSB for secondary com
        CALL   TWAIT           ;wait for data to be transferred
        CALL   IO              ;call AIC transfer routine
        CALL   TWAIT           ;wait for data to be transferred
        STI    R1,*+AR0(48h)   ;DTR = next data for AIC control
        POP    AR0             ;restore AR0
        RETS                   ;return from subroutine
;SUBROUTiNES FOR PRIMARY OR AUXILIARY INPUT
IOPRI   PUSH   R1              ;save R1
        LDI    063h,R1         ;load secondary com data into R1
        CALL   SW_IO           ;call IO routine to switch inputs
        POP    R1              ;restore R1
        RETS                   ;return from subroutine
IOAUX   PUSH   R1              ;save R1
        LDI    073h,R1         ;load secondary com data into R1
        CALL   SW_IO           ;call IO routine to switch inputs
        POP    R1              ;restore R1
        RETS                   ;return from subroutine
```

FIGURE 3.8 (continued)

procedure. The two extended-precision registers R6 and R7 are selected for input and output, respectively.

2. Assemble LOOP.ASM (not AICCOM31.ASM) and run it. Apply a sinusoidal input with an amplitude between 1 and 3 V and a frequency between 500 and 3 kHz. Verify a delayed output signal of the same frequency as the input signal.

3. To test the AIC auxiliary input AUX IN, change the fourth value in AIC-

SEC from 67h to 77h. This sets bit d4 to 1 (see Figure 3.4) and selects AUX IN, available from pin 3 of the 32-pin edge connector JP3 on the DSK board. Verify that the delayed output has the same frequency as the input but with an amplitude reduced by two.

4. Verify that the primary input IN is available from pin 1 on JP3. Note that bit d4 within the AIC control register in Figure 3.4 must be set to zero in order to access the primary input.

5. Bits d6 and d7 in the AIC control register determine the gain control. Change the fourth value 67h to 27h to set bits d6 and d7 to zero and verify that the output amplitude is reduced by two. Change 67h to 0A7 to set bit d6 to zero and bit d7 to one, and verify that the output amplitude is increased by two.

6. Bypass the AIC input bandpass filter with bit d2 in the AIC control register set to zero by changing 67h to 63h in AICSEC. Increase the input signal frequency to slightly above 4000 Hz. The output signal will appear as a signal with a lower frequency, referred to as an *aliased* signal. The input bandpass filter on the AIC removes these imaging effects. The input filter is set with a bandwidth less than the ideal Nyquist frequency, referred to as one-half the sampling frequency. Increase the input signal frequency to approximately 5 kHz, then to 9 kHz and observe these imaging effects. An aliased signal is present at 3 kHz, then at 1 kHz.

Example 3.4 Loop/Echo with Interrupt Using TMS320C3x Code

Figure 3.9 shows the loop or echo program LOOPI.ASM, which illustrates conversion rate or sampling rate using interrupt. Consider the following.

1. An interrupt service routine with the label ISR is defined within the section "intsect" which is at the address 809FC5. As shown in Figure A.3, interrupt XINT0 is selected.

2. AICSET_I and AICIO_I initialize and invoke the AIC input and output routines for interrupt. The IDLE instruction waits for an interrupt to occur. On interrupt, execution proceeds to the interrupt service routine ISR. The AIC input and output routines are then invoked with AICIO_I. Execution returns, with the return from interrupt instruction RETI. The instruction LDI R6,R7 is then executed, which loads the input from R6 into R7 for output. The branch instruction BR LOOP causes execution to return to the IDLE instruction and wait for the next interrupt to occur.

3. The AIC input bandpass filter is bypassed by using 63h in lieu of 67h in AICSEC with bit d2 = 0 in Figure 3.4. Input a sinusoidal signal and increase the input frequency beyond 4 kHz. Observe the aliasing effects as you increase the input signal frequency beyond the BW of the input filter on the AIC. Do you observe an aliased 1-kHz signal when the input signal frequency is 9 kHz? See also the previous loop program example, which uses a polling procedure to obtain an output sample rate.

```
;LOOPI.ASM - LOOP PROGRAM USING INTERRUPT
          .start    "intsect",0x809FC5 ;starting address for interrupt
          .start    ".text",0x809900   ;starting address for text
          .start    ".data",0x809C00   ;starting address for data
          .include "AICCOM31.ASM"      ;AIC communication routines
          .sect     "intsect"          ;section for interrupt vector
          BR        ISR                ;XINT0 interrupt vector
          .data                        ;data section
AICSEC    .word     162Ch,1h,4892h,63h ;Fs = 8 kHz
          .entry    BEGIN              ;start of code
          .text                        ;text section
BEGIN     LDP       AICSEC             ;init to data page 128
          CALL      AICSET_I           ;init AIC
LOOP      IDLE                         ;wait for transmit interrupt
          LDI       R6,R7              ;output R7=new input in R6
          BR        LOOP               ;branch back to LOOP
ISR       CALL      AICIO_I            ;output R7, R6=input
          RETI                         ;return from interrupt
```

FIGURE 3.9 Loop/echo program with interrupt (LOOPI.ASM).

Example 3.5 Sine Generation with Interrupt Using TMS320C3x Code

Figure 3.10 shows the program listing SINE8I.ASM, which is the interrupt-driven version of SINE4P.ASM in Example 1.2 and uses eight points to generate a sinusoid. On interrupt, execution proceeds to the interrupt service routine ISR. The first value (zero) contained in the memory address specified by AR1 is loaded into R7. When the AIC input/output routines are invoked, the output is in R7. In this example, processing for input (using R6) is not necessary. The instruction RETI causes execution to return to the IDLE instruction either directly or after the BR WAIT instruction, and waits until the next interrupt occurs.

Run this program and verify that it generates an output sinusoid with a frequency of $f = 1$ kHz, the ratio of the sampling rate and the number of points. An FM signal can be implemented based on the program SINE8I.ASM. See Experiment 3 in Section 3.7.

Example 3.6 Pseudorandom Noise Generation Using TMS320C3x Code

A 32-bit random noise sequence is generated using the following scheme shown in Figure 3.11:

a) A 32-bit seed or initial value is chosen (for example, 7E521603h).

```
;SINE8I.ASM - GENERATES A SINE WITH 8 POINTS USING INTERRUPTS
           .start    "intsect",0x809FC5  ;starting addr for interrupt
           .start    ".text",0x809900    ;starting address for text
           .start    ".data",0x809C00    ;starting address for data
           .include  "AICCOM31.ASM"      ;AIC communication routines
           .sect     "intsect"           ;section for interrupt vector
           BR        ISR                 ;XINT0 interrupt vector
           .data                         ;data section
AICSEC     .word     162Ch,1h,4892h,67h  ;Fs = 8 kHz
SINE_ADDR  .word     SINE_VAL            ;starting addr of sine values
           .brstart  "SINE_BUFF",16      ;align sine table
SINE_VAL   .word     0,707,1000,707,0,-707,-1000,-707 ;sine values
LENGTH     .set      8                   ;length of circular buffer
           .entry    BEGIN               ;start of code
           .text                         ;text section
BEGIN      LDP       AICSEC              ;init to data page 128
           CALL      AICSET_I            ;init AIC
           LDI       LENGTH,BK           ;BK=size of circular buffer
           LDI       @SINE_ADDR,AR1      ;AR1=starting addr of sine values
WAIT       IDLE                          ;wait for interrupt
           BR        WAIT                ;branch to wait until interrupt
;   INTERRUPT SERVICE ROUTINE
ISR        LDI       *AR1++%,R7          ;R7=sine value for output
           CALL      AICIO_I             ;call AIC for output
           RETI                          ;return from interrupt
```

FIGURE 3.10 Sine generation program with interrupt (SINE8I.ASM).

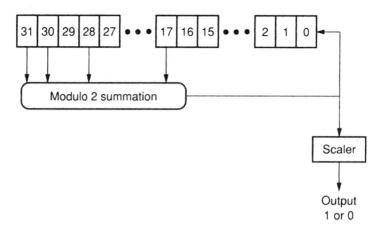

FIGURE 3.11 Pseudorandom noise generator diagram.

b) A modulo 2 sum of bits 17, 28, 30, and 31 is performed.

c) The LSB of the result (0 or 1) is tested and scaled to a positive or to a negative value.

d) The seed value is shifted left by one and the previous resulting bit is placed in the LSB position and the process repeated.

Figure 3.12 shows the program listing PRNOISE.ASM that generates a

```
;PRNOISE.ASM - PSEUDORANDOM NOISE GENERATOR
        .start   ".text",0x809900  ;starting address of text
        .start   ".data",0x809C00  ;starting address of data
        .include "AICCOM31.ASM"     ;AIC communication routines
        .data                       ;data section
AICSEC .word   162Ch,1h,4892h,67h ;Fs = 8 kHz
SEED   .word   7E521603h           ;initial seed value
PLUS   .word   400h                ;positive level
MINUS  .word   0FFFFFC00h          ;negative level
        .entry  BEGIN               ;start of code
        .text                       ;text section
BEGIN  LDP     AICSEC              ;init to data page 128
       CALL    AICSET              ;initialize AIC
       LDI     @SEED,R0            ;R0 =initial seed value
LOOP   LDI     R0,R4               ;put seed in R4
       LSH     -31,R4              ;move bit 31 to LSB    =>R4
       LDI     R0,R2               ;R2 = R0 = SEED
       LSH     -30,R2              ;move bit 30 to LSB    =>R2
       ADDI    R2,R4               ;add bits (31+30)      =>R4
       LDI     R0,R2               ;R2 = R0 = SEED
       LSH     -28,R2              ;move bit 28 to LSB    =>R2
       ADDI    R2,R4               ;add bits (31+30+28)   =>R4
       LDI     R0,R2               ;R2 = R0 = SEED
       LSH     -17,R2              ;move bit 17 to LSB    =>R2
       ADDI    R2,R4               ;add bits(31+30+28+17)=>R4
       AND     1,R4                ;mask LSB of R4
       LDIZ    @MINUS,R7           ;if R4=0, R7 = MINUS value
       LDINZ   @PLUS,R7            ;if R4=1, R7 = PLUS  value
       LSH     1,R0                ;shift new seed left by 1
       OR      R4,R0               ;put R4 into LSB of R0
       CALL    AICIO_P             ;output in R7 using AIC routine
       BR      LOOP                ;repeat for next noise sample
```

FIGURE 3.12 Pseudorandom noise generation program (PRNOISE.ASM).

pseudorandom noise, with the output rate of each noise sample determined by polling. Note the following from Figure 3.12.

1. The output sequence is scaled by 400h or by 0FFFFFC00h, defined in PLUS or MINUS in the program, which correspond to ±1024.

2. The instruction LSH -31,R4 is a logical shift and is to the right (with the minus), which brings bit 31 of the seed value into the LSB location. The selected bits are first moved into the LSB location before being summed.

3. Assemble and run this program. Connect the output to a spectrum analyzer. The shareware utility Goldwave (see Section 1.4 and Appendix B) requires a PC and a sound card to turn Goldwave into a virtual instrument as a spectrum analyzer. The output appears flat (with averaging) and rolls off at approximately 3550 Hz, the AIC input filter bandwidth.

4. The amplitude level of the noise spectrum is determined by the PLUS and MINUS scaling factors. Change PLUS and MINUS to 1000h and 0FFFFF000h, respectively, which correspond to ±4096 and verify that the amplitude spectrum is higher.

In Chapter 4, we will generate the output random noise (internally) as an input to a filter so that we can observe the characteristics and frequency response of the filter on a spectrum analyzer.

Example 3.7 Alternative Pseudorandom Noise Generation with Interrupt Using TMS320C3x Code

Figure 3.13 shows the interrupt-driven program PRNOISEI.ASM that generates the same output noise as in the previous example. Consider the following.

1. Shifting the seed value 17 locations to the right moves bit 17 to the LSB position. When the seed value is again shifted by 11, this places bit 28 into the LSB location (having already been shifted by 17). In the previous example, the original seed value is reloaded each time before it is shifted, whereas in this example it is not.

2. On interrupt, execution proceeds to the interrupt vector address or service routine specified by ISR, where each generated output sample is determined by the interrupt rate or sampling rate (even though there is no input).

3. Before the procedure for generating each noise sample is repeated, the seed value is shifted left by one, and the resulting bit (0 or 1) in R4 is placed in the LSB position of R0, which now contains the new seed value (shifted by one). After each noise sample, program execution returns to the IDLE instruction, either directly or first through the BR WAIT instruction to wait for the next interrupt to occur, and then the interrupt service routine is repeated.

4. Verify the same type of output noise as in the previous example. Use a larger sampling rate such as 20 kHz in order to obtain a wider spectrum before it rolls off. For a sampling rate of 20 kHz, the A and B registers were calculated previously in Section 3.2, where A = 0x0A14 and B = 0x3E7E.

```
;PRNOISEI.ASM - ALTERNATIVE NOISE GENERATOR USING INTERRUPT
        .start   "intsect",0x809FC5 ;starting address for interrupt
        .start   ".text",0x809900   ;starting address for text
        .start   ".data",0x809C00   ;starting address for data
        .include "AICCOM31.ASM"     ;AIC communication routines
        .sect    "intsect"          ;interrupt vector section
        BR       ISR                ;XINT0 interrupt vector
        .data                       ;data section
AICSEC .word    162Ch,1h,4892h,67h ;Fs = 8 kHz
SEED   .word    7E521603H          ;initial seed value
PLUS   .word    1000h              ;positive noise level
MINUS  .word    0FFFFF000H         ;negative noise level
        .entry   BEGIN              ;start of code
        .text                       ;text section
BEGIN  LDP      AICSEC             ;init to data page 128
       LDI      @SEED,R0           ;R0=initial seed value
       LDI      0,R7               ;init R7 (output) tO 0
       CALL     AICSET_I           ;initialize AIC
WAIT   IDLE                        ;wait for interrupt
       BR       WAIT               ;branch to WAIT
;   INTERRUPT SERVICE ROUTINE
ISR    LDI      0,R4               ;init R4=0
       LDI      R0,R2              ;put seed in R2
       LSH      -17,R2             ;move bit 17 TO LSB     =>R2
       ADDI     R2,R4              ;add bit (17)           =>R4
       LSH      -11,R2             ;move bit 28 to LSB     =>R2
       ADDI     R2,R4              ;add bits (28+17)       =>R4
       LSH      -2,R2              ;bit 30 (17+11+2)to LSB=>R2
       ADDI     R2,R4              ;add bits (30+28+17)    =>R4
       LSH      -1,R2              ;move bit 31 to LSB     =>R2
       ADDI     R2,R4              ;add bits (31+30+28+17)=>R4
       AND      1,R4               ;mask LSB of R4
       LDIZ     @MINUS,R7          ;if R4 = 0, R7 = MINUS
       LDINZ    @PLUS,R7           ;if R4 = 1, R7 = PLUS
       LSH      1,R0               ;shift new seed left by 1
       OR       R4,R0              ;put R4 into LSB of R0
       CALL     AICIO_I            ;call AIC for output in R7
       RETI                        ;return from interrupt
```

FIGURE 3.13 Alternative pseudorandom noise generation program with interrupt (PRNOISEI.ASM).

Example 3.8 Loop/Echo with AIC Data Using C Code

Figure 3.14 shows a listing of the program LOOPALL.C, which incorporates the AIC data for initialization and input/output communication using polling (see also the program SINEALL.ASM in Example 3.2). It can serve as a sample program that contains the code necessary to communicate with the AIC. The executable file LOOPALL.OUT is on disk and can be downloaded and run on the DSK. Chapter 1 describes the use of the optional tools for compiling and linking C-coded programs. Consider the following.

1. In certain situations, the optimizing C compiler reduces a repetitive statement with a variable that changes as a result of an external event such as an interrupt service routine to a single read statement. To prevent this, the volatile declaration in the program tells the compiler not to optimize, for example, any references to BPASE (Figure 3.14), since it is pointing at a peripheral address 808000. We use volatile int *PBASE as a pointer to that memory address.

2. Setting the timer period register to 1 at 808028 produces an output clock frequency of 6.25 MHz, which can be verified from pin 8 on the DSK board connector JP1.

3. When the UPDATE_SAMPLE function is called, an output occurs from the data transmit register at 808048. This output is first shifted left by two to enable primary AIC communication. Before each output and input, the transmit buffer is first cleared. An input sample is obtained from the data receive register at memory location 80804C. The input sample is sign-extended by first shifting the data left by 16 bits and then right by 18 bits. Note that the AIC has a 14-bit ADC and DAC.

In the next example, we will show a smaller program calling the AIC routines contained in a separate file.

The executable COFF file is on the accompanying disk. Input a sinusoidal signal with a frequency between 1 and 3 kHz and an amplitude between 1 and 3 V, and verify the same result as with the loop programs implemented in C3x code in Examples 3.3 and 3.4. Change the timer period register to double the AIC master clock to 12.5 MHz. This effectively doubles the sampling frequency or output rate and increases the bandwidth of the input filter on the AIC.

Example 3.9 Loop/Echo Calling AIC Routines in Separate File, Using C Code

Figure 3.15 shows a listing of the program LOOPC.C that calls the AIC communication routines contained in the program AICCOMC.C shown in Figure 3.16.

Verify the same result as in Examples 3.3, 3.4, and 3.8 yielding a delayed output sinusoid with the same frequency as an input sinusoid.

```
/*LOOPALL.C - LOOP/ECHO WITH AIC DATA INCORPORATED FOR I/O          */
#define TWAIT while (!(PBASE[0x40] & 0x2)) /*wait till XMIT buffer clear*/
int AICSEC[4]= {0x162C,0x1,0x4892,0x67};   /*config data for SP0 AIC    */
volatile int *PBASE = (volatile int *) 0x808000; /*peripherals base addr*/

void AICSET()                              /*function to initialize AIC  */
{
 volatile int loop;                        /*declare local variables     */
 PBASE[0x28] = 0x00000001;                 /*set timer period            */
 PBASE[0x20] = 0x000003C1;                 /*set timer control register  */
 asm("    LDI  00000002h,IOF");            /*set IOF low to reset AIC     */
 for (loop = 0; loop < 50; loop++);        /*keep IOF low for a while     */
 PBASE[0x42] = 0x00000131;                 /*set xmit port control       */
 PBASE[0x43] = 0x00000131;                 /*set receive port control    */
 PBASE[0x40] = 0x0E970300;                 /*set serial port global reg  */
 PBASE[0x48] = 0x00000000;                 /*clear xmit register         */
 asm("    OR   00000006h,IOF");            /*set IOF high to enable AIC   */
 for (loop = 0; loop < 4; loop++)          /*loop to configure AIC       */
   {
    TWAIT;                                 /*wait till XMIT buffer clear  */
    PBASE[0x48] = 0x3;                      /*enable secondary comm       */
    TWAIT;                                 /*wait till XMIT buffer clear  */
    PBASE[0x48] = AICSEC[loop];             /*secondary command for SP0   */
   }
}
int UPDATE_SAMPLE(int output)              /*function to update sample   */
{
  int input;                               /*declare local variables     */
  TWAIT;                                   /*wait till XMIT buffer clear  */
  PBASE[0x48] = output << 2;               /*left shift and output sample */
  input = PBASE[0x4C] << 16 >> 18;         /*input sample and sign extend */
  return(input);                           /*return new sample           */
}

main()
{
 int data_in, data_out;                    /*initialize variables        */
 AICSET();                                 /*call function to config AIC  */
 while (1)                                 /*create endless loop          */
  {
  data_in = UPDATE_SAMPLE(data_out);       /*call function to update sample*/
  data_out = data_in;                      /*loop input to output        */
  }
}
```

FIGURE 3.14 Loop program with AIC data incorporated using C code (LOOPALL.C).

```
/*LOOPC.C - LOOP PROGRAM WITH POLLING                            */
#include "aiccomc.c"                         /*AIC comm routines   */
int AICSEC[4] = {0x162C,0x1,0x4892,0x67}; /*Config data for AIC  */
main()
{
  int  data_in, data_out;                   /*init variables      */
  AICSET();                                 /*config AIC          */
  while (1)                                 /*create endless loop */
  {
   data_in = UPDATE_SAMPLE(data_out);       /*func to update sample*/
   data_out = data_in;                      /*loop input to output */
  }
}
```

FIGURE 3.15 Loop program calling AIC routines using C code (LOOPC.C).

AICCOMC.C

1. The AIC communication program AICCOMC.C is the C-coded version of the assembly coded program AICCOM31.ASM listed in Figure 3.8. It is derived from the program LOOPALL.C listed in Figure 3.14.

2. The wait loop is used to keep the IOF register low for a while (recommended by Texas Instruments). The transmit, receive, and serial port global control registers at 808042, 808043, and 808040, respectively, are initialized. The serial port data transmit register at 808048 is initialized to zero. The AIC is enabled by setting the IOF register high.

3. The AIC is ready to transmit a new word when bit 1 (XRDY) of the serial port global control register at 808040 in Figure A.6 is set to 1; otherwise, a wait loop is executed.

4. If the AIC is interrupt-driven, then AICSET_I, is accessed, which first initializes the AIC as with polling, then enables the AIC for interrupt (with the asm statements). The use of C3x code within a C program must be done carefully, such as the instructions to initialize the I/O flag (IOF), the interrupt flag (IF), the interrupt enable (IE), and the status (ST) registers with asm commands. The C3x code within these asm statements is ignored by the C compiler and is executed as specified. Note the blank space after the quotation, since an instruction must not start in column 1. The interrupt flag register is initialized to zero in order to clear any pending interrupts. After the IE register is set and the GIE bit within the status ST register is set to 1, interrupt is enabled.

5. Since the AIC has a 14-bit ADC and DAC, the output value is left-shifted by two and the LSBs are cleared to zero to enable primary communication of the AIC.

```c
/*AICCOMC.C - COMUNICATION ROUTINES FOR AIC*/
#define TWAIT while (!(PBASE[0x40] & 0x2)) /*wait till XMIT buffer clear*/
extern int AICSEC[4];                       /*array defined in main prog */
volatile int *PBASE = (volatile int *) 0x808000; /*peripherals base addr*/

void AICSET()                            /*function to initialize AIC  */
{
volatile int loop;                       /*declare local variables     */
PBASE[0x28] = 0x00000001;                /*set timer period            */
PBASE[0x20] = 0x000003C1;                /*set timer control register  */
asm("    LDI      00000062h,IOF");       /*set IOF low to reset AIC     */
for (loop = 0; loop < 90; loop++);       /*keep IOF low for a while     */
PBASE[0x42] = 0x00000131;                /*set xmit port control        */
PBASE[0x43] = 0x00000131;                /*set receive port control     */
PBASE[0x40] = 0x0E970300;                /*set serial port global reg   */
PBASE[0x48] = 0x00000000;                /*clear xmit register          */
asm("    OR       00000006h,IOF");       /*set IOF high to enable AIC   */
for (loop = 0; loop < 4; loop++)         /*loop to configure AIC        */
{
 TWAIT;                                  /*wait till XMIT buffer clear */
 PBASE[0x48] = 0x3;                      /*enable secondary comm        */
 TWAIT;                                  /*wait till XMIT buffer clear */
 PBASE[0x48] = AICSEC[loop];             /*secondary command for SP0    */
}
}

void AICSET_I()                          /*configure AIC, enable TINT0 */
{
 AICSET();                               /*function to configure AIC    */
 asm("    LDI      00000000h,IF");       /*clear IF Register            */
 asm("    OR       00000010h,IE");       /*enable EXINT0 CPU interrupt */
 asm("    OR       00002000h,ST");       /*global interrupt enable      */
}

int UPDATE_SAMPLE(int output)            /*function to update sample    */
{
 int input;                             /*declare local variables      */
 TWAIT;                                  /*wait till XMIT buffer clear */
 PBASE[0x48] = output << 2;              /*left shift and output sample*/
 input = PBASE[0x4C] << 16 >> 18;        /*input sample and sign extend*/
 return(input);                          /*return new sample            */
}
```

FIGURE 3.16 AIC communication program using C code (AICCOMC.C).

6. For input and output, the data receive register at 80804C and the data transmit register at 808048, are used.

Example 3.10 Loop/Echo with Interrupt Using C Code

Figure 3.17 is a listing of the program LOOPCI.C which calls the AIC routines in AICCOMC.C and is the interrupt-driven version of the program LOOPC.C in the previous example. It uses the interrupt structure supported by C. Consider the following.

1. Interrupt XINT0 is chosen using the transmit interrupt vector c_int05 (Figure A.3). After initializing the AIC so that it becomes interrupt-driven, execution proceeds to wait within an endless loop for the interrupt to occur.

2. On interrupt, execution proceeds to the interrupt vector function c_int05. The data transmit register is at the peripheral address 808048, where each output value is stored.

3. The program VECS_DSK.ASM is assembled with the TMS320 floating-point tools to create the object file VECS_DSK.OBJ (on the accompanying disk). It defines the interrupt address and is linked with the main program. It contains the following:

```
.ref      _c_int05      ;select XINT0
.sect     "vecs"        ;section for interrupt vectors
br        _c_int05      ;use XINT0 for interrupt
```

```
/*LOOPCI.C - LOOP PROGRAM USING INTERRUPTS               */
#include "aiccomc.c"                  /*AIC comm routines      */
int AICSEC[4] = {0x162C,0x1,0x4892,0x67}; /*AIC data for Fs = 8 kHz */
int data_in, data_out;               /*declare global variables*/


void c_int05()                       /*XINT0 interrupt routine */
{
  data_in = UPDATE_SAMPLE(data_out); /*update sample to SP0 AIC*/
  data_out = data_in;                /*loop input to output    */
}


main()
{
  AICSET_I();                        /*configure SP0 of AIC    */
  for (;;);                          /*wait for interrupt      */
}
```

FIGURE 3.17 Loop program with interrupt using C code (LOOPCI.C).

The linker command file LOOPCI.CMD is on the accompanying disk. Chapter 1 describes the use of the TMS320 floating-point tools.

4. Run this program and verify the same result as in the two previous examples.

3.5 PC HOST–TMS320C31 COMMUNICATION

Communication between the PC host and the TMS320C31 on the DSK is initiated by downloading the kernel via the PC's parallel port. Before the kernel is downloaded, the TMS320C31 is reset by toggling the INIT signal from the PC's parallel port. This causes the TMS320C31 to boot load from address 0xFFF000, which is mapped to the PC's parallel port. Once the kernel is downloaded, the communication routines are in place.

Depending on whether the parallel port is bidirectional, the actual data transferred to the DSK is either 8 bits or 4 bits. 32-bit wide data is reconstructed after several bytes or nibbles. The communication routines provided with the DSK tools determine the correct transfer (byte or nibble).

The getmem routine used by the PC to communicate with the DSK can read any block of 32-bit data accessible by the TMS320C31 address bus through the parallel port to the PC. The putmem routine used by the PC to communicate with the DSK can write any block of 32-bit data from the PC through the parallel port to any memory location accessible by the TMS320C31.

Several utilities such as TARGET.CPP, DRIVE.CPP, OBJECT.CPP, available on the disk with the software tools for the DSK, allow for communications with the C31, reads/writes a block of data from/to the C31.

The two following examples illustrate communication or interaction between the PC host and the C31 on the DSK with the functions putmem and getmem provided with the software tools. These two examples make use of the TMS320 floating-point DSP assembly language tools and Borland's C/C++ compiler, version 5.0 [7]. Two files—DSKLIB.H and DSKLIB.LIB—need to be created first.

Support Header File DSKLIB.H

The header file DSKLIB.H (on the accompanying disk) contains the following support header files:

```
#include <dos.h>
#include <bios.h>
#include <conio.h>
#include <ctype.h>
#include <stdio.h>
```

```
#include <stdlib.h>
#include <time.h>
#include <string.h>
#include "dsk.h"
#include "errormsg.h"
#include "dsk_coff"
#include "keydef.h"
```

The first eight header files are included with the Borland's C/C++ compiler and the last four header files are provided in the DSK software tools.

Library Support File DSKLIB.LIB Using Borland's C/C++ Compiler

Several utilities, such as TARGET.CPP, DRIVE.CPP, and OBJECT.CPP, provided with the DSK tools, allow for communications with the C31. It is convenient to have one library file that contains these support files for communication between the PC host and the C31. The library support file DSKLIB.LIB (on the accompanying disk) can be created using Borland's C/C++ compiler, version 5.0:

1. Select File → New → Project
2. Enter C:\DSKTOOLS\DSKLIB.IDE for the project Path and Name
3. Select Static Library [for .exe][.lib] as Target Type
4. Select DOS (Standard) and Large for Platform and Target Model, respectively. Press OK for other options as default. This creates the project file DSKLIB.IDE
5. Select DSKLIB.LIB and click with the right mouse button to Add Node and add the following nine files as nodes (clicking with right mouse button to add each node): driver.cpp, tmsfloat.cpp, textwin.cpp, target.cpp, symbols.cpp, rand386.cpp, object.cpp, errormsg.cpp, dsk_coff.cpp. These files are provided with the DSK tools.
6. Select Project → Build all to create the library support file DSKLIB.LIB

For this process, select the file symbols.h (on the accompanying disk), which is slightly different than the version provided with the DSK tools (two changes are made in symbols.h associated with typedef enum). Otherwise, two errors (need an identifier to declare) will result in compiling. Several additional files with other extensions for DSKLIB are created during this process and they can be ignored.

Example 3.11 PC—TMS320C31 Communication Using C Code

This example illustrates the use of some utilities that support communication between the PC host and the C31 on the DSK. Communication can be accomplished using the function putmem to transmit data from the PC host to the C31 and the function getmem to receive data from the C31 DSK. Figure 3.18 shows a listing of the program PCCOM.CPP that sends a number to the C31 through memory location 809800 using the function putmem(.); where (.) consists of the memory address to send the data value, how many values, and the data. The program PCCOM.CPP is compiled with the C/C++ compiler [7] to create the executable file PCCOM.EXE. Several header files that support the PC-TMS320C31 communication are included within the single header file DSKLIB.H (for convenience). This header file is "included" in the program PCCOM.CPP. The number received by the C31 (sent by the PC host) is multiplied by two and the result is sent back to the PC host through memory address 809801, as shown in the program C31COM.C, listed in Figure 3.19.

The program C31COM.C is compiled and linked with the TMS320 floating-point DSP assembly language tools (not included with the DSK package) to create the executable file C31COM.OUT (on the accompanying disk). Example 1.3 describes the use of the floating-point tools.

Compiling/Linking PCCOM.CPP

Compile the source file PCCOM.CPP with Borland's C/C++ compiler as follows:

1. Select File → New → Project
2. Enter C:\DSKTOOLS\PCCOM.IDE for the Project Path and Name
3. Select Application [.exe] as Target Type
4. Select DOS (Standard) and Large for Platform and Target Model, respectively. Press OK for other options as default. This creates the project file PCCOM.IDE
5. Select PCCOM.EXE, and click with right mouse button to Add Node, and add the file DSKLIB.LIB (created previously) as a node within the project (PCCOM.CPP is already added as a node)
6. Select Project → Build all to create the executable file PCCOM.EXE.

Execute on the PC host PCCOM.EXE and enter a value. Verify that the C31 multiplies the user's value by two. Note the following from the source file PCCOM.CPP:

1. PCCOM.EXE is executed by the PC host, which also downloads C31COM.OUT into the C31 to run.

```cpp
//PCCOM.CPP PC - TMS320C31 COMMUNICATION TO MULTIPLY TWO NUMBERS
#include "dsklib.h"                    //contains several header files
char DSK_APP[]="C31COM.OUT";
char DSK_EXE[]="PCCOM.EXE";

void config_dsk_for_comm()
{
  MSGS err; //enumerated message for looking up messages
  clrscr();
  Scan_Command_line(DSK_EXE);
  Detect_Windows();

  // Download the communications kernel
  for(;;)
  {
    if(Init_Communication(10000) == NO_ERR) break;
    if(kbhit()) exit(0);
  }
  HALT_CPU();                          //load applications code
  if((err=Load_File(DSK_APP,LOAD))!=NO_ERR)
  { printf("%s %s\n",DSK_APP,Error_Strg(err));
    exit(0);
  }
  RUN_CPU();   //DSK is initialized & able to communicate
  clrscr();
}

void main()
{
  char ch[10];
  unsigned long hostdata = 0;
  config_dsk_for_comm();
  do
  {
    clrscr();
    printf("Enter a number to transmit to C31 to be multiplied by 2\n");
    scanf ("%ld", &hostdata);
    putmem(0x809800L, 1, &hostdata);
    printf("The value written to the DSK is %ld. \n", hostdata);
    hostdata = 0;
    getmem(0x809801L, 1, &hostdata);
    printf("The value returned by the DSK is %ld. \n", hostdata);
    printf ("Press 'Y' to continue or 'Q' to Quit. \n");
    scanf("%s", &ch);
  } while (toupper(ch[0]) != 'Q');
}
```

FIGURE 3.18 PC program for communication with TMS320C31 (PCCOM.CPP).

```
/*C31COM.C - C31 COMMUNICATION PROGRAM TO MULTIPLY NUMBER BY 2*/
main()
{
 unsigned int hostdata;
 unsigned int *pwAddrHD;
 unsigned int *pwAddrTD;
 asm("      OR   2000h,ST  ");
 for (;;)
  {
   pwAddrHD = (unsigned int *)0x809800; /*PC host mail Addr   */
   pwAddrTD = (unsigned int *)0x809801; /*C31 target mail Addr*/
   hostdata = *pwAddrHD;
   *pwAddrTD = hostdata * 2;
  }
}
```

FIGURE 3.19 TMS320C31 program for communication with PC (C31COM.C).

2. The DSK is initialized and configured for communications using the function config_dsk_for_comm.

3. The Detect_Windows function determines if Windows is currently operating. If so, the multitasking feature is disabled before reading and writing data to the parallel port.

4. The Init_Communication function attempts to communicate with the DSK kernel if one exists. If not, the DSK is reset and the DSK kernel is downloaded.

5. The Halt_CPU function halts the execution of instructions in preparation for downloading the DSK executable file.

6. The Load_File function downloads the DSK file and sets the program counter. If the file is loaded successfully, the RUN_CPU function is executed, which begins executing code at the current program counter.

7. The Long format (L) associated with the addresses 0x809800 and 0x809801 is used to allow for values greater than 16,383.

A real-time loop control in the following example further illustrates these communication functions.

Example 3.12 Loop Control with PC—TMS320C31 Communication Using C Code

The utility functions putmem and getmem are illustrated to provide real-time control of a loop program. The function putmem is used to send/write to the C31 an attenuation value in order to change the amplitude of an output signal.

1. Figure 3.20 shows a listing of the program PCLOOP.CPP, compiled and linked using Borland's C/C++ compiler as in the previous example to create the executable file PCLOOP.EXE, which executes on the PC host. The program C31LOOP.C shown in Figure 3.21 is compiled using Texas Instruments' C compiler to create the object file C31LOOP.OBJ, which must be linked (see Example 1.3) to create the output executable file C31LOOP.OUT that runs on the C31 DSK.

2. Execute PCLOOP.EXE to run both programs, since it downloads and runs C31LOOP.OUT on the DSK.

3. Verify that a sinusoidal input produces a delayed output sinusoid with the same frequency as the input, but with an amplitude determined by the user's selected attenuation value. Input a sinusoidal signal with an amplitude between 1 and 3 V with a frequency between 1 and 3 kHz. Verify that a selected attenuation value of two or four decreases the output sinusoid amplitude by two or four, respectively. (See also the program LOOPC.C listed in Figure 3.15.)

Example 3.13 Data Acquisition with the DSK Using C and TMS320C3x Code

This example illustrates the capability of the DSK as a data acquisition tool using some of the support files provided with the DSK tools. A total of 512 sample points are acquired, stored into a file, and can then be processed. An example discussed in Appendix B uses MATLAB to plot and take the FFT (FFT is discussed in Chapter 6) of the acquired data, displaying on the monitor screen both the time and frequency domains of the acquired data. To implement this data acquisition example:

1. Assemble/link the file DAQ.CPP (on the accompanying disk) with Borland's C/C++, as in the previous two examples with the programs PC-COM.CPP and PCLOOP.CPP, to create the executable PC host file DAQ.EXE.

2. Assemble DAQ.ASM (on the accompanying disk) with the DSK tools to create the executable DSK file DAQ.DSK

3. Input a sinusoidal signal with a frequency of 3 kHz into the DSK.

4. Type DAQ.EXE to execute the data acquisition programs. This downloads and runs the executable DSK file DAQ.DSK and stores 512 data sample points, which represents the 3-kHz signal, into the file DAQ.DAT.

Verify that these sample points represent the 3-kHz sinusoidal signal.

```cpp
//PCLOOP.CPP - LOOP WITH AMPLITUDE CONTROL
#include "dsklib.h"          //contains several header files
char DSK_APP[] = "C31LOOP.OUT";
char DSK_EXE[] = "PCLOOP.EXE";

void config_dsk_for_comm()
{
  MSGS err; //enumerated message for looking up messages
  clrscr();
  Scan_Command_line(DSK_EXE);
  Detect_Windows();

  // Download the communications kernel
  for(;;)
  {
    if(Init_Communication(10000) == NO_ERR) break;
    if(kbhit()) exit(0);
  }
  HALT_CPU();                     //load applications code
  if((err=Load_File(DSK_APP,LOAD))!=NO_ERR)
  { printf("%s %s\n",DSK_APP,Error_Strg(err));
    exit(0);
  }
  RUN_CPU();   //DSK is initialized & able to communicate
  clrscr();
}

void main()
{
  unsigned long hostdata = 0;
  config_dsk_for_comm();  //call function to config for comm
  for(;;)
   {
    clrscr();
    printf("\n\n");
    printf("\nEnter Attenuation value (1-10) or CTRL-BREAK to quit: ");
    scanf ("%d", &hostdata);
    putmem(0x809800L, 1, &hostdata);
   }
}
```

FIGURE 3.20 PC program for real-time loop control with TMS320C31 (PCLOOP.CPP).

```
/*C31LOOP.C - LOOP PROGRAM WITH AMPLITUDE CONTROL              */
#include "aiccomc.c"                   /*AIC communications routines*/
int AICSEC[4]={0x162C,0x1,0x4892,0x67}; /*AIC setup data        */

void main(void)
{
 unsigned int *pAmpt;
 unsigned int temp;
 int data_IN, data_OUT, ampt = 1;      /*declare variables      */
 asm("    OR    2000h,ST ");
 AICSET();                             /*initialize AIC         */
 pAmpt = (unsigned int *)0x809800;
 do
   {
   temp = *pAmpt;                  /*pAmpt is pointer to value from host*/
   if (temp > 0 && temp < 11) ampt=temp; /*temp is attenuation value*/
   data_IN = UPDATE_SAMPLE(data_OUT);   /*input sample          */
   data_OUT = data_IN / ampt;          /*scale input to output  */
   }
 while (1);                            /*endless loop           */
}
```

FIGURE 3.21 TMS320C31 program for real-time loop control with PC (C31LOOP.C).

Change the input signal frequency to 2 kHz, execute DAQ.EXE again and verify that the newly acquired 512 data sample points represent the 2-kHz signal. This can be done readily with MATLAB, as described in Appendix B.

3.6 EXTERNAL/FLASH MEMORY AND I/O WITH 16-BIT STEREO AUDIO CODEC

External and Flash Memory

Although the C31 has 2K words of internal memory, the last 256 memory locations are used for the communications kernel and vectors. No additional memory is available on the DSK board. Appendix C describes a daughter board with 32K words (32-bit) of SRAM with zero wait state memory and 128K bytes of flash memory. See also references 6 and 8. This daughter board fits underneath (connects to) the DSK board through the four connectors JP2-3 and JP5-6 along the edge of the DSK board. All the necessary signals (address, data, V+, GND, R/W, INT0-3, STRB) from the C31 are available through these four connectors for use by the SRAM and the flash memory. With the external memory, application programs that require more memory space than allocated internally by the C31 can be implemented. A ten-band multirate filter that requires more memory than available with the DSK is described in Chapter 8.

Appendix C illustrates how an application-specific program can be stored on this daughter board and run without the use of a PC. A power supply similar to the one on the DSK is also on the daughter board. For example, without any PC, a filter can be implemented (run) by simply turning on the power to the daughter board, which is connected underneath the DSK board.

Alternative I/O with the Crystal CS4216/CS4218 Stereo Audio Codec

Appendix D describes a homemade board that contains a Crystal CS4216 (or CS4218) 16-bit stereo audio codec and connects to the DSK board through the connector JP1. This board contains jacks for line and microphone inputs. An evaluation board with the CS4216/CS4218 is commercially available from Crystal Semiconductor, Inc. The CS4216/CS4218 uses Delta-Sigma A/D and D/A converters with internal 64x oversampling, and internal input antialiasing and output reconstruction filters [8–10]. A maximum sampling rate of 50 kHz can be obtained. Appendix D contains some programming examples with the CS4216/CS4218.

Super DSK, commercially available from Kane Computing [11], interfaces to the DSK. It contains external and flash memories as well as a 16-bit codec for a maximum sampling rate of 50 kHz.

3.7 EXPERIMENT 3: INPUT AND OUTPUT WITH THE DSK

1. Implement Examples 3.1–3.5.

2. Verify the AIC master clock frequency from pin 8 on the DSK board JP1 connector. Verify that a timer value of 0 (from 1) in the program AICCOMC.C doubles the AIC master clock from 6.25 MHz to 12.5 MHz, which effectively doubles the sampling frequency.

3. The program SINEFM.ASM (on disk) extends the sine generator program SINE8I.ASM to implement an FM signal using 128 points. Verify a sweeping sinusoidal signal. Note that the index register IR0 specifies the step size.

4. Implement the two pseudorandom noise generation programs in Examples 3.6 and 3.7. Choose different scaling factors such as 800h and 0FFFFF800h and different sampling rates in the pseudorandom noise generator program and verify changes in the amplitude spectrum and the frequency before roll-off occurs.

5. Implement a pseudorandom noise generator using a method such that the output values are stored in consecutive memory locations. Examples 3.6 and 3.7 illustrate the noise generator algorithm. In Experiment 2 (Chapter 2, questions 3 and 4), we showed how an output sequence can be stored in consecutive memory locations as well as on disk. Scale the output sequence by ±4096 setting PLUS with 1000h and MINUS with 0FFFFF000h. Show that the output sequence is (before scaling): 1, 1, 1, 1, 0, 1, 0, 1, 1, 1, 1, 0, 0, ...

6. Implement Examples 3.8–3.10.

7. Implement Examples 3.11 and 3.12.

8. Implement a sine generator in C with four points, using interrupt (see the loop program example generated with interrupt, LOOPCI.C). Select a sampling frequency of 8 kHz. The following code can be useful:

```
int loop = 0;
int sin_table[4] = {0, ...};
```

where {0, ... } specifies the four data points (scaled) and represents the sine sequence. The interrupt function follows:

```
PBASE[0x48] = sin_table[loop] << 2;
if (loop < 3) ++loop;
else loop = 0;
```

Note that the four values stored in the sine table array are sent to the data transmit register one at a time at a rate specified by the sampling frequency. Verify an output sinusoid at a frequency of $F_s/4$.

Section 8.4 describes a project which extends this example. It uses external interrupt to control the amplitude of the generated sinewave.

REFERENCES

1. *TLC32040C, TLC32040I, TLC32041C, TLC32041I Analog Interface Circuits,* Texas Instruments, Inc., Dallas, TX, 1995.

2. R. Chassaing and D. W. Horning, *Digital Signal Processing with the TMS320C25,* Wiley, New York, 1990.

3. R. Chassaing, *Digital Signal Processing with C and the TMS320C30,* Wiley, New York, 1992.

4. *TMS320C3x DSP Starter Kit User's Guide,* Texas Instruments, Inc., Dallas, TX, 1996.

5. C. W. Solomon, "Switched-Capacitor Filters," IEEE Spectrum, June 1988.

6. *TMS320C3x User's Guide,* Texas Instruments, Inc., Dallas, TX, 1997.

7. *Borland C/C++ Compiler,* Borland International Inc., Scotts Valley, CA.

8. *TMS320C3x General-Purpose Applications User's Guide,* Texas Instruments, Inc. Dallas, TX, 1998.

9. J. C. Candy and G. C. Temes eds., *Oversampling Delta-Sigma Data Converters—Theory, Design and Simulation,* IEEE Press, New York, 1992.

10. P. M. Aziz, H. V. Sorensen, and J. Van Der Spiegel, "An Overview of Sigma Delta Converters," *IEEE Signal Processing Magazine,* Jan. 1996.

11. Super DSK, from Kane Computing, at www.kanecomputing.com/kanecomputing.

4

Finite Impulse Response Filters

- Introduction to the z-transform
- Design and implementation of finite impulse response (FIR) filters
- Programming examples using C and TMS320C3x code

The z-transform is introduced in conjunction with discrete-time signals. Mapping from the s-plane, associated with the Laplace transform, to the z-plane, associated with the z-transform, is illustrated. FIR filters are designed with the Fourier series method and implemented by programming a discrete convolution equation. Effects of window functions on the characteristics of FIR filters are covered.

4.1 INTRODUCTION TO THE z-TRANSFORM

The z-transform is utilized for the analysis of discrete-time signals, similar to the Laplace transform for continuous-time signals. We can use the Laplace transform to solve a differential equation that represents an analog filter, or the z-transform to solve a difference equation that represents a digital filter. Consider an analog signal $x(t)$ ideally sampled

$$x_s(t) = \sum_{k=0}^{\infty} x(t)\delta(t - kT) \qquad (4.1)$$

where $\delta(t - kT)$ is the impulse (delta) function delayed by kT, and $T = 1/F_s$ is the sampling period. The function $x_s(t)$ is zero everywhere except at $t = kT$. The Laplace transform of $x_s(t)$ is

91

$$X_s(s) = \int_0^\infty x_s(t) e^{-st}\, dt$$

$$= \int_0^\infty \{x(t)\delta(t) + x(t)\delta(t-T) + \ldots\} e^{-st}\, dt \tag{4.2}$$

From the property of the impulse function

$$\int_0^\infty f(t)\delta(t - kT)dt = f(kT)$$

$X_s(s)$ in (4.2) becomes

$$X_s(s) = x(0) + x(T)e^{-sT} + x(2T)e^{-2sT} + \ldots = \sum_{n=0}^\infty x(nT)e^{-nsT} \tag{4.3}$$

Let $z = e^{sT}$ in (4.3), which becomes

$$X(z) = \sum_{n=0}^\infty x(nT)z^{-n} \tag{4.4}$$

Let the sampling period T be implied; then $x(nT)$ can be written as $x(n)$, and (4.4) becomes

$$X(z) = \sum_{n=0}^\infty x(n)z^{-n} = ZT\{x(n)\} \tag{4.5}$$

which represents the z-transform (ZT) of $x(n)$. There is a one-to-one correspondence between $x(n)$ and $X(z)$, making the z-transform a unique transformation.

Exercise 4.1 *ZT* of Exponential Function $x(n) = e^{nk}$

The ZT of $x(n) = e^{nk}$, $n \geq 0$, and k a constant is

$$X(z) = \sum_{n=0}^\infty e^{nk} z^{-n} = \sum_{n=0}^\infty (e^k z^{-1})^n \tag{4.6}$$

Using the geometric series, obtained from a Taylor series approximation

$$\sum_{n=0}^\infty u^n = \frac{1}{1-u} \qquad |u| < 1$$

(4.6) becomes

$$X(z) = \frac{1}{1 - e^k z^{-1}} = \frac{z}{z - e^k} \tag{4.7}$$

for $|e^k z^{-1}| < 1$, or $|z| > |e^k|$. If $k = 0$, then the ZT of $x(n) = 1$ is $X(z) = z/(z-1)$.

Exercise 4.2 ZT of Sinusoid $x(n) = \sin n\omega T$

A sinusoidal function can be written in terms of complex exponentials. From Euler's formula $e^{ju} = \cos u + j \sin u$

$$\sin n\omega T = \frac{e^{jn\omega T} - e^{-jn\omega T}}{2j}$$

Then

$$X(z) = \frac{1}{2j} \sum_{n=0}^{\infty} \{e^{jn\omega T} z^{-n} - e^{-jn\omega T} z^{-n}\} \tag{4.8}$$

Using the geometric series as in the previous exercise, one can solve for $X(z)$; or the results in (4.7) can be used with $k = j\omega T$ in the first summation of (4.8) and $k = -j\omega T$ in the second, to yield

$$X(z) = \frac{1}{2j} \left\{ \frac{z}{z - e^{j\omega T}} - \frac{z}{z - e^{-j\omega T}} \right\}$$

$$= \frac{1}{2j} \left\{ \frac{z^2 - ze^{-j\omega T} - z^2 + ze^{j\omega T}}{z^2 - z(e^{-j\omega T} + e^{j\omega T}) + 1} \right\}$$

$$= \frac{z \sin \omega T}{z^2 - 2z \cos \omega T + 1} \tag{4.9}$$

$$= \frac{Cz}{z^2 - Az - B} \qquad |z| > 1 \tag{4.10}$$

where $A = 2 \cos \omega T$
$\qquad B = -1$
$\qquad C = \sin \omega T$

We will generate a sinusoid in Chapter 5 based on this result. We can readily generate sinusoidal waveforms of different frequencies by changing the value of ω in (4.9).

Similarly, using Euler's formula for $\cos n\omega T$ as a sum of two complex exponentials, one can find the ZT of $x(n) = \cos n\omega T = (e^{jn\omega T} + e^{-jn\omega T})/2$, as

$$X(z) = \frac{z^2 - z \cos \omega T}{z^2 - 2z \cos \omega T + 1} \qquad |z| > 1 \qquad (4.11)$$

Mapping from s-Plane to z-Plane

The Laplace transform can be used to determine the stability of a system. If the poles of a system are on the left side of the $j\omega$ axis on the s-plane, a time-decaying system response will result, yielding a stable system. If the poles are on the right side of the $j\omega$ axis, the response will grow in time, making such system unstable. Poles located on the $j\omega$ axis, or purely imaginary poles, will yield a sinusoidal response. The sinusoidal frequency is represented by the $j\omega$ axis, and $\omega = 0$ represents DC.

In a similar fashion, we can determine the stability of a system based on the location of its poles on the z-plane associated with the z-transform, since we can find corresponding regions between the s-plane and the z-plane. Since $z = e^{sT}$ and $s = \sigma + j\omega$

$$z = e^{\sigma T} e^{j\omega T} \qquad (4.12)$$

Hence, the magnitude of z is $|z| = e^{\sigma T}$ with a phase of $\theta = \omega T = 2\pi f/F_s$, where F_s is the sampling frequency. To illustrate the mapping from the s-plane to the z-plane, consider the following regions from Figure 4.1.

1. $\sigma < 0$. Poles on the left side of the $j\omega$ axis (region 2) in the s-plane represent a stable system, and (4.12) yields a magnitude of $|z| < 1$, because $e^{\sigma T} < 1$. As σ varies from $-\infty$ to 0^-, $|z|$ will vary from 0 to 1^-. Hence, poles *inside* the unit circle within region 2 in the z-plane will yield a stable system. The response of such system will be either a decaying exponential, if the poles are real, or a decaying sinusoid, if the poles are complex.

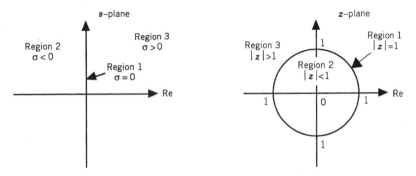

FIGURE 4.1 Mapping from s-plane to z-plane.

2. $\sigma > 0$. Poles on the right side of the $j\omega$ axis (region 3) in the s-plane represent an unstable system, and (4.12) yields a magnitude of $|z| > 1$, because $e^{\sigma T} > 1$. As σ varies from 0^+ to ∞, $|z|$ will vary from 1^+ to ∞. Hence, poles *outside* the unit circle within region 3 in the z-plane will yield an unstable system. The response of such system will be either an increasing exponential, if the poles are real, or a growing sinusoid, if the poles are complex.

3. $\sigma = 0$. Poles on the $j\omega$ axis (region 1) in the s-plane represent a marginally stable system, and (4.12) yields a magnitude of $|z| = 1$, which corresponds to region 1. Hence, poles *on* the unit circle in region 1 in the z-plane will yield a sinusoid. In Chapter 5, we will implement a digital oscillator by programming a difference equation with its poles *on* the unit circle. Note that from the previous exercise, the poles of $X(s) = \sin n\omega T$ in (4.9) or $X(s) = \cos n\omega T$ in (4.11) are the roots of $z^2 - 2z \cos \omega T + 1$, or

$$p_{1,2} = \frac{2 \cos \omega T \pm \sqrt{4 \cos^2 \omega T - 4}}{2}$$

$$= \cos \omega T \pm \sqrt{-\sin^2 \omega T} = \cos \omega T \pm j \sin \omega T \qquad (4.13)$$

The magnitude of each pole is

$$|p_1| = |p_2| = \sqrt{\cos^2 \omega T + \sin^2 \omega T} = 1 \qquad (4.14)$$

The phase of z is $\theta = \omega T = 2\pi f/F_s$. As the frequency f varies from zero to $\pm F_s/2$, the phase θ will vary from 0 to π.

Difference Equations

A digital filter is represented by a difference equation in a similar fashion as an analog filter is represented by a differential equation. To solve a difference equation, we need to find the z-transform of expressions such as $x(n - k)$, which corresponds to the kth derivative $d^k x(t)/dt^k$ of an analog signal $x(t)$. The order of the difference equation is determined by the largest value of k. For example, $k = 2$ represents a second-order derivative. From (4.5)

$$X(z) = \sum_{n=0}^{\infty} x(n)z^{-n} = x(0) + x(1)z^{-1} + x(2)z^{-2} + \ldots \qquad (4.15)$$

Then, the z-transform of $x(n - 1)$, which corresponds to a first-order derivative dx/dt is

$$ZT\{x(n - 1)\} = \sum_{n=0}^{\infty} x(n - 1)z^{-n}$$

$$= x(-1) + x(0)z^{-1} + x(1)z^{-2} + x(2)z^{-3} + \cdots$$

$$= x(-1) + z^{-1} \{ x(0) + x(1)z^{-1} + x(2)z^{-2} + \cdots \}$$

$$= x(-1) + z^{-1} X(z) \tag{4.16}$$

where we used (4.15), and $x(-1)$ represents the initial condition associated with a first-order difference equation. Similarly, the ZT of $x(n-2)$, equivalent to a second derivative $d^2 x(t)/dt^2$ is

$$ZT\{x(n-2)\} = \sum_{n=0}^{\infty} x(n-2)z^{-n}$$

$$= x(-2) + x(-1)z^{-1} + x(0)z^{-2} + x(1)z^{-3} + \cdots$$

$$= x(-2) + x(-1)z^{-1} + z^{-2} \{ x(0) + x(1)z^{-1} + \cdots \}$$

$$= x(-2) + x(-1)z^{-1} + z^{-2} X(z) \tag{4.17}$$

where $x(-2)$ and $x(-1)$ represent the two initial conditions required to solve a second-order difference equation. In general

$$ZT\{x(n-k)\} = z^{-k} \sum_{m=1}^{k} x(-m)z^{m} + z^{-k} X(z) \tag{4.18}$$

If the initial conditions are all zero, then $x(-m) = 0$ for $m = 1, 2, \ldots, k$, and (4.18) reduces to

$$ZT\{x(n-k)\} = z^{-k} X(z) \tag{4.19}$$

4.2 DISCRETE SIGNALS

A discrete signal $x(n)$ can be expressed as

$$x(n) = \sum_{m=-\infty}^{\infty} x(m)\delta(n-m) \tag{4.20}$$

where $\delta(n-m)$ is the impulse sequence $\delta(n)$ delayed by m, which is equal to one for $n = m$ and is zero otherwise. It consists of a sequence of values $x(1)$, $x(2)$, $\ldots$, where n is the time, and each sample value of the sequence is taken one sample-time apart, determined by the sampling interval or sampling period $T = 1/F_s$.

The signals and systems that we will be dealing with in this book are linear and time-invariant, where both superposition and shift invariance apply. Let an input signal $x(n)$ yield an output response $y(n)$, or $x(n) \rightarrow y(n)$. If $a_1 x_1(n) \rightarrow a_1 y_1(n)$ and $a_2 x_2(n) \rightarrow a_2 y_2(n)$, then $a_1 x_1(n) + a_2 x_2(n) \rightarrow a_1 y_1(n) + a_2 y_2(n)$, where a_1 and a_2 are constants. This is the superposition property, where an overall output response is the sum of the individual responses to each input. Shift-invariance implies that if the input is delayed by m samples, the output response will also be delayed by m samples, or $x(n - m) \rightarrow y(n - m)$. If the input is a unit impulse $\delta(n)$, the resulting output response is $h(n)$, or $\delta(n) \rightarrow h(n)$, and $h(n)$ is designated as the impulse response. A delayed impulse $\delta(n - m)$ yields the output response $h(n - m)$ by the shift-invariance property.

Furthermore, if this impulse is multiplied by $x(m)$, then $x(m)\delta(n - m) \rightarrow x(m)h(n - m)$. Using (4.20), the response becomes

$$y(n) = \sum_{m = -\infty}^{\infty} x(m)h(n - m) \qquad (4.21)$$

which represents a convolution equation. For a causal system, (4.21) becomes

$$y(n) = \sum_{m = -\infty}^{n} x(m)h(n - m) \qquad (4.22)$$

Letting $k = n - m$ in (4.22)

$$y(n) = \sum_{k = 0}^{\infty} h(k)x(n - k) \qquad (4.23)$$

4.3 FINITE IMPULSE RESPONSE FILTERS

Filtering is one of the most useful signal processing operations [1–34]. Digital signal processors are now available to implement digital filters in real-time. The TMS320C31 instruction set and architecture makes it well suited for such filtering operations. An analog filter operates on continuous signals and is typically realized with discrete components such as operational amplifiers, resistors, and capacitors. However, a digital filter, such as a finite impulse response (FIR) filter, operates on discrete-time signals and can be implemented with a digital signal processor such as the TMS320C31. This involves the use of an ADC to capture an external input signal, processing the input samples, and sending the resulting output through a DAC.

Within the last few years, the cost of digital signal processors has been significantly reduced, which adds to the numerous advantages that digital filters have over their analog counterparts. These include higher reliability, accuracy, and less sensitivity to temperature and aging. Stringent magnitude and phase characteristics can be realized with a digital filter. Filter characteristics such as center frequency, bandwidth, and filter type can be readily modified. A number of tools are available to quickly design and implement within a few minutes an FIR filter in real-time using the TMS320C31-based DSK. The filter design consists of the approximation of a transfer function with a resulting set of coefficients.

Different techniques are available for the design of FIR filters, such as a commonly used technique that utilizes the Fourier series, as discussed in the next section. Computer-aided design techniques such as that of Parks and McClellan are also used for the design of FIR filters [4–5].

The convolution equation (4.23) is very useful for the design of FIR filters, since we can approximate it with a finite number of terms, or

$$y(n) = \sum_{k=0}^{N-1} h(k)x(n-k) \tag{4.24}$$

If the input is a unit impulse $x(n) = \delta(0)$, the output impulse response will be $y(n) = h(n)$. We will see in the next section how to design an FIR filter with N coefficients $h(0), h(1), \ldots, h(N-1)$, and N input samples $x(n), x(n-1), \ldots, x(n-(N-1))$. The input sample at time n is $x(n)$, and the delayed input samples are $x(n-1), \ldots, x(n-(N-1))$. Equation (4.24) shows that an FIR filter can be implemented with the knowledge of the input $x(n)$ at time n and of the delayed inputs $x(n-k)$. It is nonrecursive and no feedback or past outputs are required. Filters with feedback (recursive) that require past outputs are discussed in Chapter 5. Other names used for FIR filters are transversal and tapped-delay filters.

The z-transform of (4.24) with zero initial conditions yields

$$Y(z) = h(0)X(z) + h(1)z^{-1}X(z) + h(2)z^{-2}X(z) + \ldots + h(N-1)z^{-(N-1)}X(z) \tag{4.25}$$

Equation (4.24) represents a convolution in time between the coefficients and the input samples, which is equivalent to a multiplication in the frequency domain, or

$$Y(z) = H(z)X(z) \tag{4.26}$$

where $H(z) = ZT\{h(k)\}$ is the transfer function, or

$$H(z) = \sum_{k=0}^{N-1} h(k)z^{-k} = h(0) + h(1)z^{-1} + h(2)z^{-2} + \ldots + h(N-1)z^{-(N-1)}$$

$$= \frac{h(0)z^{(N-1)} + h(1)z^{N-2} + h(2)z^{N-3} + \ldots + h(N-1)}{z^{-(N-1)}} \tag{4.27}$$

which shows that there are $N - 1$ poles, all of which are located at the origin. Hence, this FIR filter is inherently stable, with its poles located only inside the unit circle. We usually describe an FIR filter as a filter with "no poles." Figure 4.2 shows an FIR filter structure representing (4.24) and (4.25).

A very useful feature of an FIR filter is that it can guarantee *linear* phase. The linear phase feature can be very useful in applications such as speech analysis, where phase distortion can be very critical. For example, with linear phase, all input sinusoidal components are delayed by the same amount. Otherwise, harmonic distortion can occur.

The Fourier transform of a delayed input sample $x(n - k)$ is $e^{-j\omega kT}X(j\omega)$ yielding a phase of $\theta = -\omega kT$, which is a linear function in terms of ω. Note that the group delay function, defined as the derivative of the phase, is a constant, or $d\theta/d\omega = -kT$.

4.4 FIR IMPLEMENTATION USING FOURIER SERIES

The design of an FIR filter using a Fourier series method is such that the magnitude response of its transfer function $H(z)$ approximates a desired magnitude response. The desired transfer function is

$$H_d(\omega) = \sum_{n=-\infty}^{\infty} C_n e^{jn\omega T} \qquad |n| < \infty \tag{4.28}$$

where C_n are the Fourier series coefficients. Using a normalized frequency variable v such that $v = f/F_N$, where F_N is the Nyquist frequency, or $F_N = F_s/2$, the desired transfer function in (4.28) can be written as

$$H_d(v) = \sum_{n=-\infty}^{\infty} C_n e^{jn\pi v} \tag{4.29}$$

where $\omega T = 2\pi f/F_s = \pi v$, and $|v| < 1$. The coefficients C_n are defined as

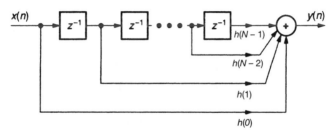

FIGURE 4.2 FIR filter structure showing delays.

$$C_n = \frac{1}{2} \int_{-1}^{1} H_d(v) e^{-jn\pi v} \, dv$$

$$= \frac{1}{2} \int_{-1}^{1} H_d(v) \{\cos n\pi v - j \sin n\pi v\} dv \tag{4.30}$$

Assume that $H_d(v)$ is an even function (frequency selective filter), then (4.30) reduces to

$$C_n = \int_{0}^{1} H_d(v) \cos n\pi v \, dv \qquad n \geq 0 \tag{4.31}$$

since $H_d(v) \sin n\pi v$ is an odd function and

$$\int_{-1}^{1} H_d(v) \sin n\pi v \, dv = 0$$

with $C_n = C_{-n}$. The desired transfer function $H_d(v)$ in (4.29) is expressed in terms of an infinite number of coefficients, and in order to obtain a realizable filter, we must truncate (4.29), which yields the approximated transfer function

$$H_a(v) = \sum_{n=-Q}^{Q} C_n e^{jn\pi v} \tag{4.32}$$

where Q is positive and finite and determines the order of the filter. The larger the value of Q, the higher the order of the FIR filter, and the better the approximation in (4.32) of the desired transfer function. The truncation of the infinite series with a finite number of terms results in ignoring the contribution of the terms outside a rectangular window function between $-Q$ and $+Q$. In the next section we will see how the characteristics of a filter can be improved by using window functions other than rectangular.

Let $z = e^{j\pi v}$, then (4.32) becomes

$$H_a(z) = \sum_{n=-Q}^{Q} C_n z^n \tag{4.33}$$

with the impulse response coefficients $C_{-Q}, C_{-Q+1}, \ldots, C_{-1}, C_0, C_1, \ldots, C_{Q-1}, C_Q$. The approximated transfer function in (4.33), with positive powers of z, implies a noncausal or not realizable filter that would produce an output before an input was applied. To remedy this situation, we introduce a delay of Q samples in (4.33) to yield

$$H(z) = z^{-Q} H_a(z) = \sum_{n=-Q}^{Q} C_n z^{n-Q} \tag{4.34}$$

Let $n - Q = -i$, then $H(z)$ in (4.34) becomes

$$H(z) = \sum_{i=0}^{2Q} C_{Q-i} z^{-i} \tag{4.35}$$

Let $h_i = C_{Q-i}$ and $N - 1 = 2Q$, then $H(z)$ becomes

$$H(z) = \sum_{i=0}^{N-1} h_i z^{-i} \tag{4.36}$$

where $H(z)$ is expressed in terms of the impulse response coefficients h_i, and $h_0 = C_Q$, $h_1 = C_{Q-1}, \ldots, h_Q = C_0$, $h_{Q+1} = C_{-1} = C_1, \ldots, h_{2Q} = C_{-Q}$. The impulse response coefficients are symmetric about h_Q, with $C_n = C_{-n}$.

The order of the filter is $N = 2Q + 1$. For example, if $Q = 5$, the filter will have 11 coefficients $h_0, h_1, \ldots, h_{10}$, or

$$h_0 = h_{10} = C_5$$
$$h_1 = h_9 = C_4$$
$$h_2 = h_8 = C_3$$
$$h_3 = h_7 = C_2$$
$$h_4 = h_6 = C_1$$
$$h_5 = C_0$$

Figure 4.3 shows the desired transfer functions $H_d(v)$ ideally represented for the frequency-selective filters: lowpass, highpass, bandpass, and bandstop for which the coefficients $C_n = C_{-n}$ can be found.

1. Lowpass. $C_0 = v_1$

$$C_n = \int_0^{v_1} H_d(v) \cos n\pi v \, dv = \frac{\sin n\pi v_1}{n\pi} \tag{4.37}$$

2. Highpass. $C_0 = 1 - v_1$

$$C_n = \sum_{v_1}^{1} H_d(v) \cos n\pi v \, dv = -\frac{\sin n\pi v_1}{n\pi} \tag{4.38}$$

3. Bandpass. $C_0 = v_2 - v_1$

$$C_n = \int_{v_1}^{v_2} H_d(v) \cos n\pi v \, dv = \frac{\sin n\pi v_2 - \sin n\pi v_1}{n\pi} \tag{4.39}$$

4. Bandstop. $C_0 = 1 - (v_2 - v_1)$

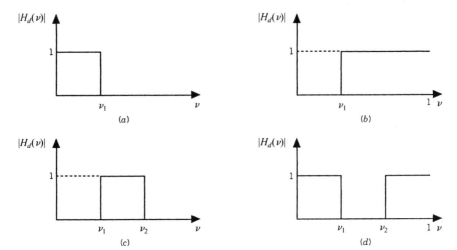

FIGURE 4.3   Desired transfer function: (a) lowpass; (b) highpass; (c) bandpass; (d) band-stop.

$$C_n = \int_0^{v_1} H_d(v)\cos n\pi v \, dv + \int_{v_2}^1 H_d(v)\cos n\pi v \, dv = \frac{\sin n\pi v_1 - \sin n\pi v_2}{n\pi} \quad (4.40)$$

where v_1 and v_2 are the normalized cutoff frequencies shown in Figure 4.3. Several filter-design packages are currently available for the design of FIR filters, as discussed later. When we implement an FIR filter, we will develop a generic program such that the specific coefficients will determine the filter type (whether it is a lowpass or a bandpass).

Exercise 4.3 Lowpass FIR Filter

We will find the impulse response coefficients of an FIR filter with $N = 11$, a sampling frequency of 10 kHz, and a cutoff frequency $f_c = 1$ kHz.
    From (4.37),

$$C_0 = v_1 = f_c/F_N = 0.2$$

where $F_N = F_s/2$ is the Nyquist frequency, and

$$C_n = \frac{\sin 0.2n\pi}{n\pi} \qquad n = \pm1, \pm2, \ldots, \pm5 \qquad (4.41)$$

Since the impulse response coefficients $h_i = C_{Q-i}$, $C_n = C_{-n}$, and $Q = 5$, the impulse response coefficients are

$$h_0 = h_{10} = 0$$
$$h_1 = h_9 = 0.0468$$
$$h_2 = h_8 = 0.1009$$
$$h_3 = h_7 = 0.1514$$
$$h_4 = h_6 = 0.1872$$
$$h_5 = 0.2 \tag{4.42}$$

These coefficients can be calculated with a utility program (on the accompanying disk) and inserted within a generic filter program, as described later. Note the symmetry of these coefficients about $Q = 5$. While $N = 11$ for an FIR filter is low for a practical design, doubling this number can yield an FIR filter with much better characteristics, such as selectivity, etc.

For an FIR filter to have linear phase, the coefficients must be symmetric as in (4.42).

4.5 WINDOW FUNCTIONS

We truncated the infinite series in the transfer function equation (4.29) to arrive at (4.32). We essentially put a rectangular window function with an amplitude of 1 between $-Q$ and $+Q$, and ignored the coefficients outside that window. The wider this rectangular window, the larger Q is and the more terms we use in (4.32) to get a better approximation of (4.29). The rectangular window function can therefore be defined as

$$w_R(n) = \begin{cases} 1 & \text{for } |n| \leq Q \\ 0 & \text{otherwise} \end{cases} \tag{4.43}$$

The transform of the rectangular window function $\omega_R(n)$ yields a sinc function in the frequency domain. It can be shown that

$$W_R(v) = \sum_{n=-Q}^{Q} e^{jn\pi v} = e^{-jQ\pi v} \left[\sum_{n=0}^{2Q} e^{jn\pi v} \right] = \frac{\sin\left[\left(\dfrac{2Q+1}{2} \right) \pi v \right]}{\sin(\pi v/2)} \tag{4.44}$$

which is a sinc function that exhibits high sidelobes or oscillations caused by the abrupt truncation, specifically, near discontinuities.

A number of window functions are currently available to reduce these high-amplitude oscillations; they provide a more gradual truncation to the infinite series expansion. However, while these alternative window functions reduce the amplitude of the sidelobes, they also have a wider mainlobe, which results in a filter with lower selectivity. A measure of a filter's performance is a ripple factor that compares the peak of the first sidelobe to the peak of the main lobe (their ratio). A compromise or trade-off is to select a window function that can reduce the sidelobes while approaching the selectivity that can be achieved with the rectangular window function. The width of the mainlobe can be reduced by increasing the width of the window (order of the filter). We will later plot the magnitude response of an FIR filter that shows the undesirable sidelobes.

In general, the Fourier series coefficients can be written as

$$C'_n = C_n w(n) \tag{4.45}$$

where $w(n)$ is the window function. In the case of the rectangular window function, $C'_n = C_n$. The transfer function in (4.36) can then be written as

$$H'(z) = \sum_{i=0}^{N-1} h'_i z^{-i} \tag{4.46}$$

where

$$h'_i = C'_{Q-i} \qquad 0 \leq i \leq 2Q \tag{4.47}$$

The rectangular window has its highest sidelobe level down by only -13 dB from the peak of its mainlobe, resulting in oscillations with an amplitude of considerable size. On the other hand, it has the narrowest mainlobe that can provide high selectivity.

The following window functions are commonly used in the design of FIR filters [9].

Hamming Window

The Hamming window function [9,24] is

$$w_H(n) = \begin{cases} 0.54 + 0.46 \cos(n\pi/Q) & \text{for } |n| \leq Q \\ 0 & \text{otherwise} \end{cases} \tag{4.48}$$

which has the highest or first sidelobe level at approximately -43 dB from the peak of the main lobe.

Hanning Window

The Hanning or raised cosine window function is

$$w_{HA}(n) = \begin{cases} 0.5 + 0.5 \cos(n\pi/Q) & \text{for } |n| \leq Q \\ 0 & \text{otherwise} \end{cases} \qquad (4.49)$$

which has the highest or first sidelobe level at approximately -31 dB from the peak of the mainlobe.

Blackman Window

The Blackman window function is

$$w_B(n) = \begin{cases} 0.42 + 0.5 \cos(n\pi/Q) + 0.08 \cos(2n\pi/Q) & |n| \leq Q \\ 0 & \text{otherwise} \end{cases} \qquad (4.50)$$

which has the highest sidelobe level down to approximately -58 dB from the peak of the mainlobe. While the Blackman window produces the largest reduction in the sidelobe compared with the previous window functions, it has the widest mainlobe. As with the previous windows, the width of the mainlobe can be decreased by increasing the width of the window.

Kaiser Window

The design of FIR filters with the Kaiser window has become very popular in recent years. It has a variable parameter to control the size of the sidelobe with respect to the mainlobe. The Kaiser window function is

$$w_K(n) = \begin{cases} I_0(b)/I_0(a) & |n| \leq Q \\ 0 & \text{otherwise} \end{cases} \qquad (4.51)$$

where a is an empirically determined variable, and $b = a[1 - (n/Q)^2]^{1/2}$. $I_0(x)$ is the modified Bessel function of the first kind defined by

$$I_0(x) = 1 + \frac{0.25x^2}{(1!)^2} + \frac{(0.25x^2)^2}{(2!)^2} + \ldots = 1 + \sum_{n=1}^{\infty} \left[\frac{(x/2)^n}{n!} \right]^2 \qquad (4.52)$$

which converges rapidly. A trade-off between the size of the sidelobe and the width of the mainlobe can be achieved by changing the length of the window and the parameter a.

Computer-Aided Approximation

An efficient technique is the computer-aided iterative design based on the Remez exchange algorithm, which produces equiripple approximation of FIR filters [4–5]. The order of the filter and the edges of both passbands and stopbands are fixed, and the coefficients are varied to provide this equiripple approximation. This minimizes the ripple in both the passbands and the stopbands. The transition regions are left unconstrained and are considered as "don't care" regions, where the solution may fail. Several commercial filter design packages include the Parks–McClellan algorithm for the design of an FIR filter.

4.6 FILTER DESIGN PACKAGES

Within minutes, an FIR filter can be designed and implemented in real-time. Several filter design packages are available to design FIR filters, described in Appendix B:

1. The DigiFilter from DSPlus, which supports the TMS320C31 DSK
2. MATLAB from The Math Works [35]
3. From Hyperception, which includes utilities for plotting, spectral analysis, etc. [36].
4. A "homemade" package (on the accompanying disk), which calculates the coefficients using the rectangular, Hamming, Hanning, Blackman, and Kaiser windows.

4.7 PROGRAMMING EXAMPLES USING TMS320C3X AND C CODE

Several examples illustrate the implementation of FIR filters using both C and TMS320C3x code. This includes a C program calling a filter function in TMS320C3x code. Utility packages for filter design will be introduced.

The convolution equation in (4.24) is used to program and implement these filters, or

$$y(n) = \sum_{k=0}^{N-1} h(k)x(n-k)$$

$$= h(N-1)x(n-(N-1)) + h(N-2)x(n-(N-2))$$

$$+ \ldots + h(1)x(n-1) + h(0)x(n) \tag{4.53}$$

where the order of the summation is reversed. We can arrange the impulse response coefficients within a buffer in memory starting (lower-memory address) with the last coefficient $h(N-1)$. The first coefficient $h(0)$ will reside at the "bottom" of the buffer (higher-memory address) as shown in Table 4.1. The memory organization for the input samples is also shown in Table 4.1. A circu-

TABLE 4.1   TMS320C31 memory organization for convolution

	Input Samples		
Coefficients	Time n	Time $n + 1$	Time $n + 2$
AR0 → $h(N-1)$	AR1 → $x(n-(N-1))$	newest → $x(n+1)$	$x(n+1)$
$h(N-2)$	$x(n-(N-2))$	AR1 → $x(n-(N-2))$	newest → $x(n+2)$
$h(N-3)$	$x(n-(N-3))$	$x(n-(N-3))$	AR1 → $x(n-(N-3))$
⋮	⋮	⋮	⋮
$h(1)$	$x(n-1)$	$x(n-1)$	$x(n-1)$
$h(0)$	newest → $x(n)$	$x(n)$	$x(n)$

lar buffer is reserved for these samples, with the newest sample $x(n)$ at time n at the "bottom" memory location and the oldest sample $x(n-(N-1))$ at the "top" or starting address of the samples buffer. While we can also use a circular buffer for the coefficients, it is not necessary.

Initially, all the input samples $x(n)$, $x(n-1)$, . . . are set to zero. We start at time n, acquire the first sample $x(n)$ through an ADC converter and place it at the bottom (higher-memory address) of the samples buffer as shown in Table 4.1. We can do so by storing this sample $x(n)$ at time n into a memory location, whose address is specified by an auxiliary register such as AR1. AR1 will then be incremented to point at the top (lower-memory address) of the circular buffer. The same scheme was used with the program FIR4.ASM in Example 2.3. We can now multiply $h(N-1)$, the content in memory pointed by AR0, by $x(n-(N-1))$, the content in memory pointed by AR1, and accumulate. We then postincrement the two auxiliary registers AR0 and AR1 to multiply $h(N-2)x(n-(N-2))$, which is the second term in (4.53), and continue this process within a loop.

After the last multiplication at time n, AR1 is postincremented to point at the top memory address of the samples buffer where a newly acquired sample $x(n+1)$, representing the newest sample at time $n+1$, is stored next. AR1 is then postincremented to point at the memory location which contains the sample $x(n-(N-2))$, as shown in Table 4.1. The output at time $n+1$ in (4.53) then becomes

time $n + 1$:

$$y(n+1) = h(N-1)x(n-(N-2)) + h(N-2)x(n-(N-3))$$
$$+ \ldots + h(1)x(n) + h(0)x(n+1) \tag{4.54}$$

The above process is repeated to implement (4.54) for time $n+1$. The first multiplication consists of $h(N-1)x(n-(N-2))$, since AR1 is initially pointing at $x(n-(N-2))$ at time $n+1$. Similarly, at time $n+2$, (4.53) becomes time $n + 2$:

$$y(n + 2) = h(N - 1)x(n -(N - 3)) + h(N -2)x(n -(N - 4))$$
$$+ \ldots + h(2)x(n) + h(1)x(n + 1) + h(0)x(n + 2) \qquad (4.55)$$

Note that for each time n, $n + 1$, ... the last multiply operation is between $h(0)$ and the newest sample, which is $h(0)x(k)$ at time k.

Example 4.1 FIR Lowpass Filter Simulation with 11 Coefficients Using TMS320C3x Code

Figure 4.4 shows a listing of the program LP11SIM.ASM which implements a lowpass FIR filter with the 11 coefficients calculated in Exercise 4.3. A more practical FIR filter, with sharper characteristics, requires more coefficients; however, this example is instructive, since it incorporates these same coefficients. Furthermore, other types of filters with different sets of coefficients can be readily implemented with the same program. The program FIR4.ASM in Example 2.3 provides much background for this example. Assemble and run the program LP11SIM.ASM and verify the following.

1. INB and OUTB are the starting addresses of the input and output buffers, respectively. The input represents an impulse with a value of 10,000 at $n = 0$, and zero otherwise. A circular buffer XN_BUFF is created starting at the address XN, aligned on a 16-word boundary, and initialized to zero.

2. The data page is initialized to page 128. The special register BK is loaded with 11, the actual size of the circular buffer. AR1 is loaded with the bottom address of that buffer, where the first input sample value is to be stored. AR2 and AR3 are loaded with the starting addresses of the input and output buffers, respectively.

3. The filter's routine starts at the label or address LOOP and ends with the instruction STI R7, *AR3++, and is executed 11 times (repeated 10 times). The instruction DBNZD AR4, LOOP decrements the loop counter AR4 and specifies a branch with delay based on the "not zero" condition on AR4. Hence, the three subsequent instructions are executed before branching occurs.

4. The first input sample value of 10,000 is stored into the bottom memory location of the circular buffer, at 809c3a. The starting addresses of the buffers are listed at the end of the executable file LP11SIM.DSK. From the symbol reference table, INB, OUTB, and XN_BUFF start at 809c0f, 809c1a, and 809c30, respectively; and the length b (in hex) of the circular buffer is specified from the output section.

5. For each time $n = 0, 1, \ldots, 10$, the filter routine is executed. Within this filter routine, the discrete convolution equation (4.53), for each specific time n,

```
;LP11SIM.ASM - FIR LOWPASS FILTER WITH 11 COEFF FOR SIMULATION
          .start    ".text",0x809900 ;where text begins
          .start    ".data",0x809C00 ;where data begins
          .data                      ;data section
IN_ADDR  .word    INB              ;starting address for input
OUT_ADDR .word    OUTB             ;starting address for output
XB_ADDR  .word    XN+LENGTH-1      ;bottom address of circ buffer
HN_ADDR  .word    COEFF            ;starting addr of coefficients
COEFF    .float   0               ;H10
         .float   0.0468, 0.1009, 0.1514, 0.1872, 0.2
         .float   0.1872, 0.1514, 0.1009, 0.0468
H0       .float   0               ;H0
LENGTH   .set     H0-COEFF+1      ;# of coefficients
INB      .float   10000, 0, 0, 0, 0, 0, 0, 0, 0, 0, 0
OUTB     .float   0, 0, 0, 0, 0, 0, 0, 0, 0, 0, 0
         .brstart "XN_BUFF",16     ;align samples buffer
XN       .sect    "XN_BUFF"        ;section for input samples
         .loop    LENGTH           ;buffer size for samples
         .float   0.0              ;initialize samples to zero
         .endloop                  ;end of loop
         .entry   BEGIN            ;start of code
         .text
BEGIN    LDP      @0x809800        ;init to data page 128
         LDI      LENGTH,BK        ;size of circular buffer
         LDI      @XB_ADDR,AR1     ;last sample address ->AR1
         LDI      @IN_ADDR,AR2     ;input address      ->AR2
         LDI      @OUT_ADDR,AR3    ;output address     ->AR3
FILT     LDI      LENGTH-1,AR4     ;length in AR4
LOOP     LDF      *AR2++,R3        ;input new sample
         STF      R3,*AR1++%       ;store newest sample
         LDI      @HN_ADDR,AR0     ;AR0 points to H(N-1)
         LDF      0,R0             ;init R0
         LDF      0,R2             ;init R2
         RPTS     LENGTH-1         ;repeat LENGTH-1 times
         MPYF3    *AR0++,*AR1++%,R0 ;R0 = HN*XN
||       ADDF3    R0,R2,R2         ;accumulation in R2
         DBNZD    AR4,LOOP         ;delayed branch until AR4<0
         ADDF     R0,R2            ;last mult result accumulated
         FIX      R2,R7            ;convert float R2 to integer R7
         STI      R7,*AR3++        ;store into output buffer
WAIT     BR       WAIT             ;wait
         .end                      ;end
```

FIGURE 4.4 FIR lowpass filter program for simulation (LP11SIM.ASM).

is achieved with the multiply (MPYF3) and accumulate (ADDF3) instructions, which are executed 11 times. For $n = 0$, the resulting output $y(0)$ is contained in R2. For a specific n, the parallel instruction ADDF3 is executed 11 times, whereas the ADDF R0,R2 instruction is executed only once to accumulate the last product. The output value is then converted from floating-point to integer format, then stored into the output buffer.

6. Within the debugger, the command memd OUTB (or memd 0x809c1a) displays the 11 output values starting at the address OUTB which is at 809c1a. While debugger commands are not case-sensitive, the address OUTB is. The command

```
save LP11SIM.DAT,OUTB,11,L
```

saves on disk the 11 output values in decimal (ASCII Long). Verify that they represent the impulse response 0, 468, 1009, ..., 468, 0, which are the coefficients scaled by 10,000. Figure 4.5 shows a plot of the output frequency response using a utility package from Hyperception [36], with a sampling rate of 10 kHz.

The next example illustrates an FIR bandpass filter with 45 coefficients modifying slightly the program LP11SIM.ASM.

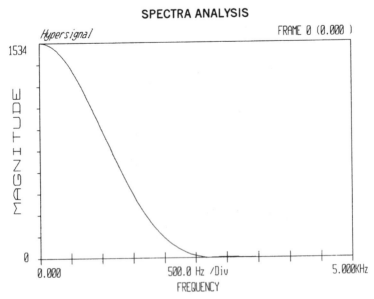

FIGURE 4.5 Frequency response of simulated FIR lowpass filter.

Example 4.2 FIR Bandpass Filter Simulation with 45 Coefficients Using TMS320C3x Code

This example makes use of the previous program example to implement a 45-coefficient FIR bandpass filter designed so that the center frequency is at $F_s/10$. The coefficients, included in the file BP45.COF and listed in Figure 4.6, were calculated with a filter design package [36] using a Kaiser window. To implement this filter, make the following changes to LP11SIM.ASM to create BP45SIM.ASM (on the accompanying disk).

1. Include the coefficient file BP45.COF in Figure 4.6 with an assembler directive. Note that the order or length (LENGTH) is already defined within the file BP45.COF.

2. Increase the length of the input values to 45 and initialize the output buffer also of length 45 to zero.

3. Align the circular buffer on a 64-word boundary.

Assemble and run this new program BP45SIM.ASM. Verify that the resulting 45 output values are –19, –27, –1, 31, 25, . . . , –19, which are the impulse response coefficients scaled by 10,000. The starting address of the output buffer is OUTB or 809c5e, as can be found at the end of the executable file BP45SIM.DSK. The output frequency response is plotted in Figure 4.7 using a sampling rate of 10 kHz. Does the circular buffer start at 809cc0 and is it of length $2d$ (45 decimal)?

In Appendix B, a slighly different version of this program is implemented to illustrate the debugger Code Explorer, an abridged version of the debugger Code Composer. It uses 64 sample points instead of 45. Within the debugger en-

```
;BP45.COF - FIR BANDPASS COEFFICIENTS (Fc = Fs/10)
        .data                           ;data section
COEFF   .float   -1.839E-3              ;H44
        .float   -2.657E-3,-1.437E-7, 3.154E-3, 2.595E-3,-4.159E-3,-1.540E-2
        .float   -2.507E-2,-2.547E-2,-1.179E-2, 1.392E-2, 4.206E-2, 5.888E-2
        .float    5.307E-2, 2.225E-2,-2.410E-2,-6.754E-2,-8.831E-2,-7.475E-2
        .float   -2.956E-2, 3.030E-2, 8.050E-2, 1.000E-1, 8.050E-2, 3.030E-2
        .float   -2.956E-2,-7.475E-2,-8.831E-2,-6.754E-2,-2.410E-2, 2.225E-2
        .float    5.307E-2, 5.888E-2, 4.206E-2, 1.392E-2,-1.179E-2,-2.547E-2
        .float   -2.507E-2,-1.540E-2,-4.159E-3, 2.595E-3, 3.154E-3,-1.437E-7
        .float   -2.657E-3              ;H1
H0      .float   -1.839E-3              ;H0
LENGTH .set      H0-COEFF+1             ;# of coefficients
```

FIGURE 4.6 Coefficient file for FIR bandpass filter (BP45.COF).

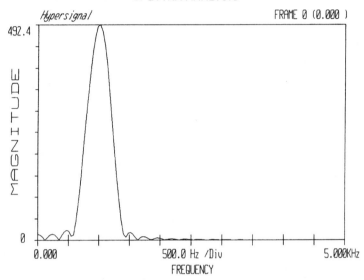

FIGURE 4.7 Frequency response of simulated FIR bandpass filter.

vironment, the output sequence can be plotted in both the time and the frequency domains.

Example 4.3 Generic FIR Filter Specified at Run-Time, Using TMS320C3x Code

Figure 4.8 shows a listing of a generic FIR filter program FIRNC.ASM which is quite similar to the two previous filter programs. The AIC communication routines are contained in the file AICCOM31.ASM and are invoked for a real-time implementation. Include the coefficient file BP45.COF (change FIR.COF to BP45.COF). Assemble and run this program, and verify that it implements a bandpass filter centered at $f = F_s/10$. Note the following.

1. The sampling rate, specified in AICSEC, is 10 kHz with the A and B registers set at 162Ch and 3872h, respectively, as discussed in Chapter 3.

2. Input a sinusoidal signal with a frequency of 1 kHz and an amplitude of less than 3 V, and observe the output signal of the same frequency. As the input signal frequency is slightly increased or decreased, the amplitude of the output signal decreases. Hence, the output signal has a maximum amplitude at $F_s/10$.

Shareware utilities such as Goldwave can be used as a virtual instrument with a PC and a sound card. While the C31 on the DSK is running, "two copies" of Goldwave can be accessed and run simultaneously as a function generator to provide a sinusoidal (or random noise) signal and as an oscilloscope (or spec-

```
;FIRNC.ASM - GENERIC FIR FILTER. INCLUDE COEFFICIENTS FILE
          .start    ".text",0x809900    ;starting address of text
          .start    ".data",0x809C00    ;starting address of data
          .include  "AICCOM31.ASM"      ;AIC communication routines
          .include  "FIR.COF"           ;coefficients file
          .data                         ;data section
XB_ADDR  .word     XN+LENGTH-1          ;last (bottom) sample address
HN_ADDR  .word     COEFF               ;starting addr of coefficients
AICSEC   .word     162ch,1h,3872h,67h  ;AIC configuration data
          .brstart  "XN_BUFF",64        ;align samples buffer
XN        .sect     "XN_BUFF"           ;buffer section for samples
          .loop     LENGTH              ;loop LENGTH times
          .float    0                   ;init samples buffer to zero
          .endloop                      ;end of loop
          .text                         ;text section
          .entry    BEGIN               ;start of code
BEGIN     LDP       AICSEC              ;init to data page 128
          CALL      AICSET              ;init AIC
          LDI       LENGTH,BK           ;BK=size of circular buffer
          LDI       @XB_ADDR,AR1        ;AR1=last sample address
FILT      LDI       LENGTH-1,AR4        ;AR4=length-1 as loop counter
LOOP      CALL      AICIO_P             ;AICIO routine,IN->R6 OUT->R7
          FLOAT     R6,R3               ;input new sample ->R3
          STF       R3,*AR1++%          ;store newest sample
          LDI       @HN_ADDR,AR0        ;AR0 points to H(N-1)
          LDF       0,R0                ;init R0
          LDF       0,R2                ;init R2
          RPTS      LENGTH-1            ;repeat LENGTH-1 times
          MPYF3     *AR0++,*AR1++%,R0   ;R0 = HN*XN
||        ADDF3     R0,R2,R2            ;accumulation in R2
          DBNZD     AR4,LOOP            ;delayed branch until AR4<0
          ADDF      R0,R2               ;last accumulation
          FIX       R2,R7               ;convert float R2 to integer R7
          NOP                           ;added due to delayed branch
          BR        FILT                ;branch to filter routine
          .end                          ;end
```

FIGURE 4.8 Generic FIR filter program (FIRNC.ASM).

trum analyzer), using Windows 95 to test this filter. The output of the sound card, with Goldwave running as a function generator, becomes the input to the DSK, while the DSK output becomes the input to the sound card, with Goldwave running as an oscilloscope or spectrum analyzer.

Coefficients Included with Batch File

Several types of filters can be readily implemented by including different sets of coefficients using a batch file. Create the following batch file FIR.BAT with the following commands:

```
copy %1 FIR.COF

DSK3A FIRNC.ASM

DSK3LOAD FIRNC.DSK
```

Use the original program FIRNC.ASM with the assembler directive .include "FIR.COF." Type:

```
FIR BP45.COF
```

This invokes the batch file FIR.BAT, copies the coefficient file BP45.COF as FIR.COF, so that FIRNC.ASM now includes the coefficient file BP45.COF. Then, FIRNC.ASM is assembled and the resulting executable file FIRNC.DSK is downloaded and run with the boot loader. Verify again that an FIR bandpass filter is implemented, centered at 1 kHz.

Several other coefficients files are included on the accompanying disk:

1. LP11.COF represents a lowpass with cutoff frequency at $F_s/10$, with 11 coefficients.
2. BP45.COF, BP33.COF, BP23.COF represent bandpass with a center frequency at $F_s/10$, and with 45, 33, and 23 coefficients, respectively.
3. BP41.COF represents a bandpass with 41 coefficients centered at $F_s/4$.
4. LP55.COF, HP55.COF, BP55.COF, and BS55.COF represent a lowpass, a highpass, a bandpass, and a bandstop, respectively; each with 55 coefficients with cutoff or center frequencies at $F_s/4$.
5. PASS2B.COF, PASS3B.COF, and PASS4B.COF represent bandpass with 2, 3, or 4 bands, respectively; each with 55 coefficients.
6. STOP3B.COF represents a bandstop with 55 coefficients and three stop bands.
7. COMB14.COF represents a comb filter with 14 coefficients.

Chapter 8 describes a program FIRALL.ASM that contains eight sets of 55 coefficients. Run the batch file FIRALL.BAT on disk. This executes a C program that prompts the user to select and run a desired filter (lowpass, highpass, etc.). Edit the file FIRALL.ASM and observe that the eight sets of coefficients are the same as the coefficients in the files listed in steps 4–6.

Example 4.4 FIR Filter Incorporating Pseudorandom Noise as Input, Using TMS320C3x Code

Figure 4.9 shows an FIR filter program FIRPRN.ASM, which incorporates the pseudorandom noise generator in Example 3.5, using interrupt (see PRNOISEI.ASM). Each output noise sample, scaled by PLUS or MINUS as ±4,096, is loaded in R7 within the noise generator routine. This noise sample is converted from floating-point to integer with the instruction FLOAT R7,R3 within the filter routine section, then stored in the memory address specified by AR1. AR1 was initially loaded with XB_ADDR, the bottom address of the circular buffer

```
;FIRPRN.ASM - FIR FILTER WITH INPUT NOISE GENERATOR USING INTERRUPT
            .start    "intsect",0x809FC5 ;starting address for interrupt
            .start    ".text",0x809900   ;starting address for text
            .start    ".data",0x809C00   ;starting address for data
            .include "AICCOM31.ASM"      ;include AIC comm routines
            .include "PASS2B.COF"        ;include coefficients file
            .sect     "intsect"          ;section for interrupt vector
            BR        ISR                ;XINT0 interrupt vector
            .data                        ;data section
AICSEC  .word     162Ch,1h,4892h,67h ;Fs = 8 KHz
SEED    .word     7E521603H            ;initial seed value
PLUS    .word     1000h                ;positive noise level
MINUS   .word     0FFFFF1000H          ;negative noise level
XB_ADDR .word     XN+LENGTH-1          ;last (bottom) sample address
HN_ADDR .word     COEFF                ;starting addr of coefficients
        .brstart "XN_BUFF",64          ;align samples buffer
XN      .sect     "XN_BUFF"            ;buffer section for samples
        .loop     LENGTH               ;buffer size of samples
        .float    0                    ;initialize samples to zero
        .endloop                       ;end of loop
```

(continued on next page)

FIGURE 4.9 FIR program incorporating input pseudorandom noise (FIRPRN.ASM).

```
          .entry   BEGIN                ;start of code
          .text                         ;text section
BEGIN     LDP      AICSEC               ;init to data page 128
          CALL     AICSET_I             ;init AIC
          LDI      @SEED,R0             ;R0=initial seed value
          LDI      0,R7                 ;init R7 (output) tO 0
          LDI      @XB_ADDR,AR1         ;last sample address => AR1
          LDI      LENGTH,BK            ;BK= length
WAIT      IDLE                          ;wait for interrupt
          BR       WAIT                 ;branch to WAIT til interrupt
;INTERRUPT SERVICE ROUTINE FOR NOISE GENERATION
ISR       LDI      0,R4                 ;init R4=0
          LDI      R0,R2                ;put seed in R2
          LSH      -17,R2               ;move bit 17 TO LSB     =>R2
          ADDI     R2,R4                ;add bit (17)           =>R4
          LSH      -11,R2               ;move bit 28 to LSB     =>R2
          ADDI     R2,R4                ;add bits (28+17)       =>R4
          LSH      -2,R2                ;move bit 30 to LSB     =>R2
          ADDI     R2,R4                ;add bits (30+28+17)    =>R4
          LSH      -1,R2                ;move bit 31 to LSB     =>R2
          ADDI     R2,R4                ;add bits (31+30+28+17) =>R4
          AND      1,R4                 ;mask LSB of R4
          LDIZ     @MINUS,R7            ;if R4 = 0, R7 = @MINUS
          LDINZ    @PLUS,R7             ;if R4 = 1, R7 = @PLUS
          LSH      1,R0                 ;shift seed left by 1
          OR       R4,R0                ;put R4 into LSB of R0
;MAIN SECTION FOR FILTER
          FLOAT    R7,R3                ;input noise sample
          STF      R3,*AR1++%           ;store newest sample
          LDI      @HN_ADDR,AR0         ;AR0 points to H(N-1)
          LDF      0,R1                 ;init R1
          LDF      0,R3                 ;init R3
          RPTS     LENGTH-1             ;repeat LENGTH-1 times
          MPYF3    *AR0++,*AR1++%,R1    ;R1 = HN*XN
||        ADDF3    R1,R3,R3             ;accumulation in R3
          ADDF3    R1,R3,R3             ;last accumulation
          FIX      R3,R7                ;convert float R3 integer R7
          CALL     AICIO_I              ;call AICIO for output
          RETI                          ;return from interrupt
          .end                          ;end
```

FIGURE 4.9   *(continued)*

reserved for the input samples. As a result, a random noise sequence internally generated becomes the input to the filter routine.

Run this program with no input connected to the DSK and verify the output frequency response of an FIR filter with two passbands as shown in Figure 4.10. The previous example lists a number of files that contain coefficients for different types of FIR filters. Use (include) the file BS55.COF, which contains the coefficients representing a bandstop FIR filter centered at $F_s/4$, and verify the frequency response shown in Figure 4.11 (obtained with an HP signal analyzer).

Example 4.5 Mixed-Mode FIR Filter With Main C Program Calling Filter Function in TMS320C3x Code

A mixed-coded implementation provides the best compromise between a C program, which is more portable and maintainable, and an equivalent assembly program, which executes faster. A C program can provide the capability for communication between the PC host and the C31, for graphical output on the PC monitor screen, etc. [26]. The C program FIRMC.C listed in Figure 4.12 calls the TMS320C3x FIR filter function FIRMCF.ASM listed in Figure 4.13. The filter function contains the time-critical section of the overall implementation and can be considered as a "black box" called from C. Examples 2.5 and 2.6, with mixed-coded programs, provide much background for this example. Note the following within the main C program FIRMC.C.

1. Since it is difficult to align a circular buffer in a data section defined in C, the buffer for the samples is chosen to be 2*N*, twice the actual size of the coefficients. This is done to ensure that any adjacent data are not accidentally overwritten.

2. The coefficient file bp45coef.h is "included" within the main C program, and shown in Figure 4.14. It represents a 45-coefficient FIR bandpass filter centered at $F_s/10$, the same set of coefficients used previously. Note that the filter's length $N = 45$ and is defined within the coefficient file.

3. The AIC is initialized and the input/output rate is set using interrupt. The C program FIRMC.C calls the TMS320C3x coded filter function filt, passing to this function the addresses of the coefficients and the data samples, the input and output addresses, and the filter's length *N*.

4. Since the function is a C-identifier, it is referenced with an underscore (_filt) within the program FIRMCF.ASM. This assembly function is derived from the generic TMS320C3x FIR program discussed previously.

Note the following within the assembly filter function.

a. The frame pointer FP set in auxiliary register AR3 is used for passing the addresses of the arguments from the main C program to the assembly function. The old frame pointer is at the first location on the stack.

b. The auxiliary registers AR4, AR5, and AR6 are dedicated, and because they are used, they must be saved using the PUSH instruction and later restored

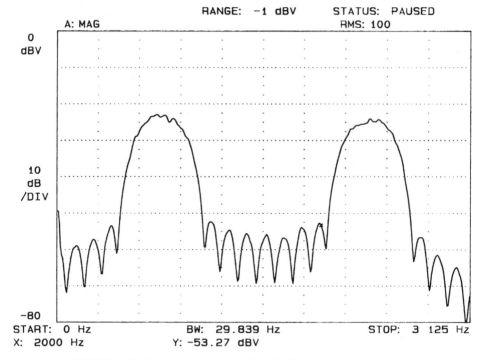

FIGURE 4.10 Frequency response of an FIR filter with two passbands.

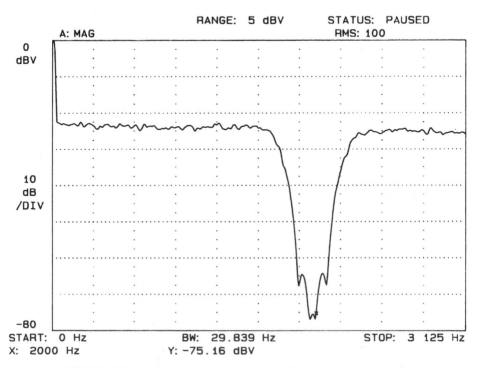

FIGURE 4.11 Frequency response of a 55-coefficient FIR bandstop filter.

```
/*FIRMC.C - FIR WITH MIXED-CODE. CALLS FUNCTION IN FIRMCF.ASM  */
#include "aiccomc.c"                    /*include AIC com routines */
#include "bp45coef.h"                   /*include coefficients file*/
float DLY[2*N];                         /*delay samples           */
int AICSEC[4] = {0x162C,0x1,0x3872,0x67}; /*AIC data, Fs=10 kHz*/
int data_in, data_out;
extern void filt(float *, float *, int *, int *, int);


void c_int05()
{
  PBASE[0x48] = data_out << 2;
  data_in = PBASE[0x4C] << 16 >> 18;
}


main ()
{
  int *IO_INPUT, *IO_OUTPUT;
  IO_INPUT = &data_in;
  IO_OUTPUT = &data_out;
  AICSET_I();
  for (;;)
    filt((float *)H, (float *)DLY, (int *)IO_INPUT, (int *)IO_OUTPUT, N);
}
```

FIGURE 4.12 C program calling TMS320C3x-coded FIR filter function (FIRMC.C).

using the POP instruction. The frame pointer also must be saved and restored. The stack pointer is loaded into the frame pointer FP.

c. The frame pointer, starting with an offset of –2, is used to point at the starting addresses of the coefficients and the input samples, the input and output addresses, and to specify the filter's length. These arguments are specified in the main C program.

Download and run FIRMC.OUT (on disk) directly into the DSK. Verify a bandpass filter centered at $F_s/10 = 1$ kHz.

If the programs are changed, the TMS320 floating-point tools described in Chapter 1 need to be used. Assemble the function FIRMCF.ASM to create the object file FIRMCF.OBJ. Compile/assemble the C program FIRMC.C to create the object file FIRMC.OBJ. This object file is then linked with the assembly function FIRMCF.OBJ and the file VECS_DSK.OBJ, which contains the interrupt definition. These files need to be assembled first with the TMS320 floating-point tools in order to create the object files. Invoke the linker command file FIRMC.CMD (on disk) to create the executable COFF file FIRMC.OUT.

```
*FIRMCF.ASM - FIR FUNCTION IN ASSEMBLY CALLED FROM FIRMC.C
FP      .set    AR3                   ;frame pointer in AR3
        .global _filt                 ;global ref/def filter routine
_filt   PUSH    FP                    ;save frame pointer FP
        LDI     SP,FP                 ;load stack pointer into FP
        PUSH    AR4                   ;save AR4
        PUSH    AR5                   ;save AR5
        PUSH    AR6                   ;save AR6
        LDI     *-FP(2),AR0           ;address of HN pointer->AR0
        LDI     *-FP(3),AR1           ;address of XN pointer->AR1
        LDI     *-FP(4),AR5           ;addr of IO_INPUT pointer->AR5
        LDI     *-FP(5),AR6           ;addr of IO_OUTPUT pointer->AR6
        LDI     *-FP(6),AR2           ;filter length ->AR2
        LDI     AR2,BK                ;size of circular buffer->BK
        SUBI    1,AR2                 ;decrement AR2
        ADDI    AR2,AR1               ;AR1=XN ADDR+LENGTH-1(BOTTOM)
        LDI     AR2,AR4               ;AR4 is loop counter
LOOP    IDLE                          ;wait for interrupt
        FLOAT   *AR5,R3               ;input new sample
        STF     R3,*AR1++%            ;store newest sample
        LDI     *-FP(2),AR0           ;AR0 points to coeff H(N-1)
        LDF     0,R0                  ;init R0
        LDF     0,R2                  ;init R2
        RPTS    AR2                   ;repeat LENGTH-1 times
        MPYF    *AR0++,*AR1++%,R0     ;R0 = HN*XN
||      ADDF    R0,R2                 ;R2 = accumulator
        DBNZD   AR4,LOOP              ;delayed branch until AR4<0
        ADDF    R0,R2                 ;last value accumulated
        FIX     R2,R0                 ;float in R2 to integer in R0
        STI     R0,*AR6               ;output R0 to IO_OUTPUT
        POP     AR6                   ;restore contents of AR6
        POP     AR5                   ;restore contents of AR5
        POP     AR4                   ;restore contents of AR4
        POP     FP                    ;restore frame pointer
        RETS                          ;return to C program
```

FIGURE 4.13 FIR filter function in TMS320C3x code called from C (FIRMCF.ASM).

```
/*BP45COEF.H-HEADER FILE COEFF FOR BANDPASS FILTER USED BY FILT*/
#define N 45                         /*length of impulse response*/
const float H[N] = {/* filter coefficients*/
-1.839E-03,-2.657E-03,-4.312E-10, 3.154E-03, 2.595E-03,-4.159E-03,
-1.540E-02,-2.507E-02,-2.547E-02,-1.179E-02, 1.392E-02, 4.206E-02,
 5.888E-02, 5.307E-02, 2.225E-02,-2.410E-02,-6.754E-02,-8.831E-02,
-7.475E-02,-2.956E-02, 3.030E-02, 8.050E-02, 1.000E-01, 8.050E-02,
 3.030E-02,-2.956E-02,-7.475E-02,-8.831E-02,-6.754E-02,-2.410E-02,
 2.225E-02, 5.307E-02, 5.888E-02, 4.206E-02, 1.392E-02,-1.179E-02,
-2.547E-02,-2.507E-02,-1.540E-02,-4.159E-03, 2.595E-03, 3.154E-03,
-4.312E-10,-2.657E-03,-1.839E-03};
```

FIGURE 4.14 Coefficient file for FIR bandpass filter (BP45COEF.H).

Example 4.6 FIR Filter With Data Move Using C Code

The previous program examples illustrate the usefulness of a circular buffer in moving the auxiliary registers or pointers in a manner that allows for the implementation of the discrete convolution equation (4.24), updating the data samples as illustrated in Table 4.1. This example illustrates a different method to implement the same convolution equation by moving the data in lieu of the pointers in order to update the data samples. Such procedure is quite common in fixed-point processors which do not have a circular buffer feature in hardware [33]. Figure 4.15 shows a listing of the program FIRDMOVE.C that incorporates this data-move procedure. Note the following:

   1. The coefficient file bp45coef.h represents the same FIR bandpass filter centered at $F_s/10$ as in the previous example, and is included for this implementation. To implement a different filter, generate a different coefficient file and "include" it in the C program. The filter's length N is specified within the coefficient file. No additional changes are necessary in the C program.

   2. The idle instruction with the asm assembly command is to wait for an interrupt. When an interrupt occurs, execution proceeds to the interrupt function c_int05. The convolution equation is achieved with the instruction

$$\text{acc+ = h[i] * dly[i];}$$

   3. To update the delay samples, a data move type of instruction is used, or

$$\text{dly[i] = dly[i - 1];}$$

Download and run the executable file FIRDMOVE.OUT (on the accompanying disk). Verify a bandpass filter centered at $F_s/10$, or 1 kHz.

```
/*FIRDMOVE.C - FIR FILTER MOVING THE DATA AND NOT POINTERS*/
#include "aiccomc.c"      /*include AIC com routines     */
#include "bp45coef.h"     /*include coefficient file     */
float DLY[N];             /*delay samples                */
int data_in, data_out;
int AICSEC[4]={0x162c,0x1,0x3872,0x67};  /*AIC config data*/

void filt(float *h,float *dly,int *IO_input,int *IO_output,int n)
{
  int i, t;
  float acc;
  for (t = 0; t < n; t++)
  {
    asm("          idle");              /*wait for interrupt */
    acc = 0.0;
    dly[0] = *IO_input;                 /*newest input sample*/
    for (i = 0; i < n; i++)
      acc += h[i] * dly[i];
    for (i = n-1; i > 0; i—)
      dly[i] = dly[i-1];               /*update samples     */
    *IO_output = acc;
  }
}


void c_int05()
{
  data_in = UPDATE_SAMPLE(data_out);
}


main ()
{
int *IO_INPUT, *IO_OUTPUT;
IO_INPUT = &data_in;
IO_OUTPUT = &data_out;
AICSET_I();
for(;;)
 filt((float *)H,(float *)DLY,(int *)IO_INPUT,(int *)IO_OUTPUT,N);
}
```

FIGURE 4.15 FIR filter program with data move using C code (FIRDMOVE.C).

Example 4.7 FIR Filter Using C Code

The program FIRC.C listed in Figure 4.16 implements an FIR filter using C code. The same coefficient file bp45coef.h that represents a bandpass filter centered at $F_s/10$ is included as in the previous example. Note the following.

1. The idle instruction within the asm assembly command is used to wait for an interrupt. When an interrupt occurs, execution proceeds to the interrupt function c_int05.

2. Within the processing loop, operations such as n − m are defined as n_m before the two loops i and j in order to increase the execution speed. The two loops i and j achieve the effects of a circular buffer to implement the convolution equation in (4.24), as illustrated in Table 4.2.

3. From the interrupt function c_int05, the UPDATE_SAMPLE function in AICCOMC.C (included) is called. Because this C function UPDATE_SAMPLE is called from another C function (c_int05), the C compiler creates a number of PUSH/PUSHF and POP/POPF, which slow down the execution time of the filter. To increase the execution speed, the following commands from UPDATE_SAMPLE can be incorporated directly:

```
PBASE[0x48] = data_out << 2;

data_in = PBASE[0x4C] << 16 >> 18;
```

The executable file FIRC.OUT (on disk) can be downloaded directly into the DSK and run. Verify that this program implements an FIR filter centered at $F_s/10$, and yields the same results as in the previous example. A different filter coefficient can be included to implement a different filter. However, if any changes are made in the source program, it needs to be recompiled/assembled and linked. Chapter 1 illustrates the use of the TMS320 floating-point tools. Although this program executes slower than the previous mixed-code implementation, it is more portable.

TABLE 4.2 Convolution for FIR filter example

	Time $n = 0$	Time $n = 1$	Time $n = 2$
	$h(0)x(44)$	$h(0)x(43)$	$h(0)x(42)$
	$h(1)x(43)$	$h(1)x(42)$	$h(1)x(41)$
	$h(2)x(42)$	$h(2)x(41)$	$h(2)x(40)$
loop i	⋮	⋮	⋮
	$h(44)x(0)$	$h(43)x(0)$	$h(42)x(0)$
loop j	—	$h(44)x(44)$	$h(43)x(44)$
			$h(44)x(43)$

```
/*FIRC.C - REAL-TIME FIR FILTER. CALLS AICCOMC.C              */
#include "aiccomc.c"                    /*include AIC com routines */
#include "bp45coef.h"                   /*include coefficients file*/
float DLY[N];                           /*delay samples            */
int data_in, data_out;
int AICSEC[4] = {0x162c,0x1,0x3872,0x67};  /*AIC data Fs =8 kHz*/

void filt(float *h,float *dly,int *IO_input,int *IO_output,int n)
{
  int i, j, m, N1, N1_m, n_m, index = 0;
  float acc = 0;
  N1 = n-1;
  dly[0] = *IO_input;
  for (m = 0; m < n; m++)
  {
    asm("    idle");                    /*wait for interrupt       */
    N1_m = N1-m;
    n_m = n-m;
    for (i = 0; i < n_m; i++)      /*addr below new sample to 0*/
      acc += h[i] * dly[N1_m-i];
    for (j = m; j > 0; j—)         /*from n to latest sample   */
      acc += h[n-j] * dly[N1_m+j]; /*latest sample last        */
    *IO_output = acc;                   /*output result            */
    acc = 0.0;                          /*clear accumulator        */
    dly[N1_m] = *IO_input;              /*get new sample           */
  }
}


void c_int05()
{
  data_in = UPDATE_SAMPLE(data_out);
}


main ()
{
 int *IO_INPUT, *IO_OUTPUT;
 IO_INPUT = &data_in;
 IO_OUTPUT = &data_out;
 AICSET_I();
 for(;;)
   filt((float *)H,(float *)DLY,(int *)IO_INPUT,(int *)IO_OUTPUT,N);
}
```

FIGURE 4.16 FIR filter program using C code (FIRC.C).

Example 4.8 FIR Filter With Samples Shifted, Using C Code

This example implements the same FIR bandpass filter as in the previous example, using a different method. It uses an array size of $2N - 1$ for the sample delays. Figure 4.17 shows a listing of the program `FIRERIC.C` that includes the same coefficient file `bp45coef.h` as in the previous example. A brief description follows.

1. Initially, the memory locations for the samples are shown in Table 4.3 (a). The last (bottom) memory location is reserved for the newest sample $x(n + 1)$ represented by `dly[44]`. This newest sample is acquired with the instruction `dly[i+k] = *IO_input`. This instruction is within a loop to acquire the newest sample for each subsequent time n.

2. The convolution operation is performed to obtain an output sample at time $n + 1$, or

$$y(n + 1) = h(0)\texttt{dly[0]} + h(1)\texttt{dly[1]} + \ldots + h(44)\texttt{dly[44]}$$

where `dly[0]` represents the oldest sample $x(n - 43)$ and `dly[44]` represents the newest sample $x(n + 1)$ as shown in Table 4.3 (a).

3. A new sample $x(n + 2)$ is then acquired and placed at the "bottom" memory location of the samples buffer following $x(n + 1)$. This newest sample is represented by `dly[45]`. The convolution operation is performed to obtain an output sample at time $n + 2$, or

$$y(n + 2) = h(0)\texttt{dly[1]} + h(1)\texttt{dly[2]} + \ldots + h(44)\texttt{dly[45]}$$

since the sample delays are updated.

TABLE 4.3 Assignment of delay samples: (a) initially; (b) after $n = 45$; (c) with process repeated

(a)	(b)	(c)
`dly[0]` → $x(n - 43)$	`dly[0]` → $x(n - 43)$	`dly[0]` → $x(n + 2)$
`dly[1]` → $x(n - 42)$	`dly[1]` → $x(n - 42)$	`dly[1]` → $x(n + 3)$
`dly[2]` → $x(n - 41)$	`dly[2]` → $x(n - 41)$	`dly[2]` → $x(n + 4)$
⋮	⋮	⋮
`dly[42]` → $x(n - 1)$	`dly[42]` → $x(n - 1)$	`dly[42]` → $x(n + 44)$
`dly[43]` → $x(n)$	`dly[43]` → $x(n)$	`dly[43]` → $x(n + 45)$
`dly[44]` → newest sample	`dly[44]` → $x(n + 1)$	`dly[44]` → $x(n + 1)$
	`dly[45]` → $x(n + 2)$	
	`dly[46]` → $x(n + 3)$	
	`dly[47]` → $x(n + 4)$	
	⋮	
	`dly[86]` → $x(n + 43)$	
	`dly[87]` → $x(n + 44)$	
	`dly[88]` → $x(n + 45)$	

```
/*FIRERIC.C - FIR FILTER WITH SAMPLES SHIFTED          */
#include "aiccomc.c"           /*include AIC com routines  */
#include "bp45coef.h"          /*include coefficient file  */
#define N 45                   /*length of impulse response*/
float DLY[N*2-1];              /*init for 2*N-1 samples    */
int data_in, data_out;
int AICSEC[4]={0x162C,0x1,0x3872,0x67}; /*AIC config data*/

void filt(float *h,float *dly,int *IO_input,int *IO_output,int n)
{
  float acc=0.0;                 /*init accumulator          */
  int   i,j,k=n-1;               /*index variables           */
  for (i = 0; i<N*2-1; i++) DLY[i] = 0.0;  /*init samples*/
  for (i=0;i<n;i++)
  {
    asm ("        idle");        /*wait for interrupt        */
    dly[i+k] =*IO_input;         /*get new sample            */
    for (j=0;j<n;j++)
       acc += h[j]*dly[i+j];     /*perform convolution       */
    *IO_output=acc;              /*output new value          */
    acc=0.0;                     /*reset accumulator         */
  }
    for (i=0;i<k;i++)            /*shift values from         */
       dly[i]=dly[i+n];         /*lower half to upper half   */
}

void c_int05()
{
data_in = UPDATE_SAMPLE(data_out);
}

main()
{
int *IO_INPUT, *IO_OUTPUT;
IO_INPUT = &data_in;
IO_OUTPUT = &data_out;
AICSET_I();
for(;;)
 filt((float *)H,(float *)DLY,(int*)IO_INPUT,(int *)IO_OUTPUT,N);
}
```

FIGURE 4.17 FIR filter program with samples shifted (FIRERIC.C).

4. For each time n, a new sample is acquired and placed at the "bottom" of the memory buffer. After 45 samples, the memory locations for the samples buffer are as shown in Table 4.3 (b), with the upper-half of the table values as in (a). Note that the size of the buffer is $2N - 1 = 89$.

5. The code within the last loop in the program shifts the samples in the lower-half memory locations to the upper-half locations as shown in Table 4.3 (c). This is equivalent to the way the samples were displayed initially, except that now the older samples are not zero. This shift is necessary for a real-time implementation where this process is continually repeated.

Run this program and verify similar results (somewhat degraded) as in the previous implementations.

Example 4.9 FIR Filter Design Using Filter Development Package

A noncommercial filter development package (FDP) is on the accompanying disk. The program FIRPROG.BAS, written in BASIC, calculates the coefficients of an FIR filter. It allows for the design of lowpass, highpass, bandpass, and bandstop FIR filters using the rectangular, Hanning, Hamming, Blackman, and Kaiser window functions. The resulting coefficients are generated in a format that contains the .float assembler directive preceding each coefficient value. The filter's length is also calculated and set within the coefficient file. This file needs to be only slightly modified and incorporated into one of the previous programs.

1. Run BASIC, and load/run the program FIRPROG.BAS. Figure 4.18 (a) and (b) show a display of available window functions and the frequency-selective filters that can be designed. Select the Kaiser window option, and a bandpass filter. A separate module for the Kaiser window is called from FIRPROG.BAS.

2. Enter the specifications shown in Figure 4.18 (c): 6 db for a passband ripple, 30 db for a stopband attenuation, etc. Then, choose f and c31 as in Figure 4.18 (d) to save the 53 resulting coefficients into a file with a format that includes the .float assembler directive. Save it as KBP53.COF.

3. Repeat the above procedure for a rectangular window. Enter 900 and 1100 Hz for the lower and upper cutoff frequencies, and 5.2 msec for the duration of the impulse response, since the number of coefficients N is

$$N = (D \times F_s) + 1$$

This will yield a design with 53 coefficients. Save the resulting coefficient file as RBP53.COF.

4. Add a .data assembler directive in each of the two files with the coefficients for the Kaiser and the rectangular windows.

5. Use the FIR filter program FIRPRN.ASM, with internally generated pseudorandom noise as input to the filter and a sampling frequency of 10 kHz.

```
                        Main Menu
                        ---------

                        1....RECTANGULAR
                        2....HANNING
                        3....HAMMING
                        4....BLACKMAN
                        5....KAISER
                        6....Exit to DOS

        Enter window desired (number only) --> 5

                        (a)

                Selections:
                        1....LOWPASS
                        2....HIGHPASS
                        3....BANDPASS
                        4....BANDSTOP
                        5....Exit back to Main Menu

        Enter desired filter type (number only) --> 3

                        (b)

        Specifications:
                        BANDPASS
                        Passband Ripple (AP) = 6 db
                        Stopband Attenuation (AS) = 30 db
                        Lower Passband Frequency = 900 Hz
                        Upper Passband Frequency = 1100 Hz
                        Lower Stopband Frequency = 600 Hz
                        Upper Stopband Frequency = 1400 Hz
                        Sampling Frequency (Fs) = 10000 Hz


        The calculated # of coefficients required is: 53

        Enter # of coefficients desired ONLY if greater than 53
        otherwise, press <Enter> to continue -->

                        (c)

Send coefficients to:
                        (S)creen
                        (P)rinter
                        (F)ile: contains TMS320 (C25 or C31) data format
                        (R)eturn to Filter Type Menu
                        (E)xit to DOS

Enter desired path --> f

Enter DSP type (C25 OR C31):? c31

                        (d)
```

FIGURE 4.18 FIR filter design with filter development package FDP (on disk): (a) choice of windows; (b) type of filter; (c) example of filter specifications; (d) menu for coefficients format.

Verify the results in Figure 4.19, which shows a plot of the frequency responses of the 53-coefficient FIR bandpass filters centered at $F_s/10$, using both the rectangular and the Kaiser window functions. The rectangular window provides a high selectivity with a sharp transition between the passbands and the stopbands. However, note how the peak of the first sidelobe is relatively high with the rectangular window, compared to the peak of the mainlobe.

4.8 EXPERIMENT 4: FIR FILTER IMPLEMENTATION

1. Test different filter characteristics with the program `FIRPRN.ASM`. Test the coefficient files `LP55.COF` and `HP55.COF`, which represent a lowpass and a highpass filter, respectively, with cutoff frequencies at $F_s/4$.

2. The frequency response of a comb filter is shown in Figure 4.20. The coefficient file `comb14.cof` on the accompanying disk contains 14 coefficients: 1, 0, 0, . . . , −1. The first and the last coefficients are not zero. An FIR comb filter contains multiple notches or stop bands that can be useful to eliminate unwanted harmonics. In such case, the harmonics would be notched out. Verify the

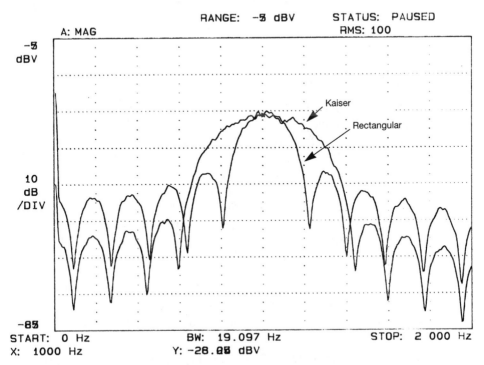

FIGURE 4.19 Frequency responses of 53-coefficient FIR filters using the rectangular and the Kaiser windows.

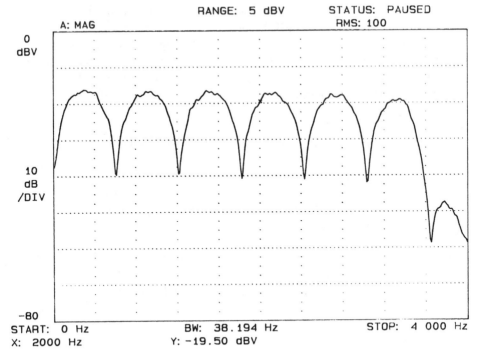

FIGURE 4.20 Frequency response of an FIR comb filter.

result with the program FIRPRN.ASM. Discuss the transfer function of such comb filter in terms of the poles and zeros in the z-plane. Reduce the number of coefficients to 10 (delete four zeros) and verify that the number of notched frequencies or stop bands are reduced.

3. Design a 37-coefficient FIR bandpass filter using an available filter design package such as MATLAB from The Math Works [35], Hypersignal from Hyperception [36], Digifilter from DSPlus (illustrated in Appendix B), or the design package FDP on the accompanying disk (as illustrated in the previous example). Choose a Kaiser window, a sampling frequency of 8 kHz, and a center frequency of 2 kHz. Implement this filter in real-time using either:

a) A TMS320C3x coded program such as FIRNC.ASM, which requires an input source, or the program FIRPRN.ASM, which incorporates a pseudorandom noise source (within the program)as input to the filter routine.

b) A C coded program such as FIRC.C or the C program FIRMC.C, which calls the assembly filter function FIRMCF.ASM. This assumes that you have access to the TMS320 floating-point tools described in Chapter 1 for compiling/assembling and linking.

4. Develop a complete FIR program BP37I.ASM that incorporates the 37-coefficients obtained in the previous question. Make it interrupt-driven to achieve a sampling rate of 8 kHz. The two programs FIRNC.ASM and FIR-PRN.ASM can be helpful.

   a) Initialize the AIC before the filter routine with AICSET_I in lieu of AICSET
   b) Use an IDLE instruction at the beginning of the filter's loop (address LOOP)
   c) Your interrupt service routine may be similar to the following:

```
ISR   CALL   AICIO_I   ;AIC I/O routines with interrupt
      RETI             ;return from interrupt
```

5. Design and implement a multiband FIR filter with 3 passbands and 65 coefficients, using a sampling frequency of 10 kHz. Each passband should be centered at a frequency of 750, 1750, and 2750 Hz, respectively. Test your results with one of the filter programs such as FIRPRN.ASM. Note that with a buffer size of 65, a circular buffer should be aligned on a 128-word boundary. Appendix B contains an example of an FIR filter design with two passbands. Figure 4.10 shows the frequency response of a 55-coefficient FIR filter with two passbands using the coefficient file PASS2B.COF in FIRPRN.ASM, with a sampling frequency of 8 kHz.

REFERENCES

1. A. V. Oppenheim and R. Schafer, *Discrete-Time Signal Processing,* Prentice-Hall, Englewood Cliffs, NJ, 1989.
2. B. Gold and C. M. Rader, *Digital Signal Processing of Signals,* McGraw-Hill, New York, 1969.
3. L. R. Rabiner and B. Gold, *Theory and Application of Digital Signal Processing,* Prentice-Hall, Englewood Cliffs, NJ, 1975.
4. T. W. Parks and J. H. McClellan, "Chebychev Approximation For Nonrecursive Digital Filter with Linear Phase," *IEEE Trans. Circuit Theory,* CT-19, 189–194 (1972).
5. J. H. McClellan and T. W. Parks, "A Unified Approach to the Design of Optimum Linear Phase Digital Filters," *IEEE Trans. Circuit Theory,* CT-20, 697–701 (1973).
6. J. F. Kaiser, "Nonrecursive Digital Filter Design Using the I_0-Sinh Window Function," in *Proceedings of the IEEE International Symposium on Circuits and Systems,* 1974.

7. J. F. Kaiser, "Some Practical Considerations in the Realization of Linear Digital Filters, in *Proceedings of the Third Allerton Conference on Circuit System Theory,* October 1965, pp. 621–633.

8. L. B. Jackson, *Digital Filters and Signal Processing,* Kluwer Academic, Norwell, MA, 1996.

9. F. J. Harris, "On the Use of Windows for Harmonic Analysis with the Discrete Fourier Transform," *Proc. IEEE,* 66, 51–83 (1978).

10. T. W. Parks and C. S. Burrus, *Digital Filter Design,* Wiley, New York, 1987.

11. S. D. Stearns and R. A. David, *Signal Processing in Fortran and C,* Prentice-Hall, Englewood Cliffs, NJ, 1993.

12. B. Porat, *A Course in Digital Signal Processing,* Wiley, New York, 1997.

13. N. Ahmed and T. Natarajan, *Discrete-Time Signals and Systems,* Reston, Reston, VA, 1983.

14. A. Antoniou, *Digital Filters: Analysis, Design, and Applications,* McGraw-Hill, New York, 1993.

15. E. C. Ifeachor and B. W. Jervis, *Digital Signal Processing A Practical Approach,* Addison-Wesley, 1993.

16. J. G. Proakis and D. G. Manolakis, *Digital Signal Processing Principles, Algorithms, and Applications,* Prentice-Hall, Englewood Cliffs, NJ, 1996.

17. D. J. DeFatta, J. G. Lucas, and W. S. Hodgkiss, *Digital Signal Processing: A System Approach,* Wiley, New York, 1988.

18. R. G. Lyons, *Understanding Digital Signal Processing,* Addison-Wesley, 1997.

19. P. A. Lynn and W. Fuerst, *Introductory Digital Signal Processing With Computer Applications,* Wiley, New York, 1994.

20. S. J. Orfanidis, *Introduction to Signal Processing,* Prentice-Hall, Englewood Cliffs, NJ, 1996.

21. A. Bateman and W. Yates, *Digital Signal Processing Design,* Computer Science Press, New York, 1991.

22. R. D. Strum and D. E. Kirk, *First Principles of Discrete Systems and Digital Signal Processing,* Addison-Wesley, Reading, MA, 1988.

23. C. S. Williams, *Designing Digital Filters,* Prentice-Hall, Englewood Cliffs, NJ, 1986.

24. R. W. Hamming, *Digital Filters,* Prentice-Hall, Englewood Cliffs, NJ, 1983.

25. S. K. Mitra and J. F. Kaiser eds., *Handbook for Digital Signal Processing,* Wiley, New York, 1993.

26. R. Chassaing, *Digital Signal Processing with C and the TMS320C30,* Wiley, New York, 1992.

27. R. Chassaing, B. Bitler, and D. W. Horning, "Real-time digital filters in C," in *Proceedings of the 1991 ASEE Annual Conference,* June 1991.

28. R. Chassaing and P. Martin, "Digital Filtering with the floating-point TMS320C30 digital signal processor," in *Proceedings of the 21st Annual Pittsburgh Conference on Modeling and Simulation,* May 1990.

29. S. K. Mitra, *Digital Signal Processing A Computer-Based Approach,* McGraw-Hill, New York, 1998.

30. R. A. Roberts and C. T. Mullis, *Digital Signal Processing,* Addison-Wesley, Reading, MA, 1987.

31. E. P. Cunningham, *Digital Filtering: An Introduction,* Houghton Mifflin, MA, 1992.

32. N.J. Loy, *An Engineer's Guide to FIR Digital Filters,* Prentice Hall, Englewood Cliffs, NJ, 1988.

33. R. Chassaing and D.W. Horning, *Digital Signal Processing with the TMS320C25,* Wiley, New York, 1990.

34. A. H. Nuttall, "Some Windows with Very Good Sidelobe Behavior," *IEEE Trans. on Acoustics, Speech, and Signal Processing,* ASSP-29, No. 1, February 1981.

35. MATLAB, The Math Works Inc., MA, 1997.

36. Hypersignal-Plus DSP Software, Hyperception, Inc., Dallas, TX, 1991.

37. TMS320C3x General-Purpose Applications User's Guide, Texas Instruments, Inc., Dallas, TX, 1998.

5

Infinite Impulse Response Filters

- Infinite impulse response filter structures: direct form I, direct form II, cascade, and parallel
- Bilinear transformation for filter design
- Sinusoidal waveform generation using difference equation
- Filter design and utility packages
- Programming examples using TMS320C3x and C code

The finite impulse response (FIR) filter discussed in the previous chapter has no analog counterpart. In this chapter, we discuss the infinite impulse response (IIR) filter that makes use of the vast knowledge already acquired with analog filters. The design procedure involves the conversion of an analog filter to an equivalent discrete filter using the bilinear transformation (BLT) technique. As such, the BLT procedure converts a transfer function of an analog filter in the s-domain into an equivalent discrete-time transfer function in the z-domain.

5.1 INTRODUCTION

Consider a general input-output equation of the form,

$$y(n) = \sum_{k=0}^{N} a_k x(n-k) - \sum_{j=1}^{M} b_j y(n-j) \tag{5.1}$$

$$= a_0 x(n) + a_1 x(n-1) + a_2 x(n-2) + \ldots + a_N x(n-N)$$
$$- b_1 y(n-1) - b_2 y(n-2) - \ldots - b_M y(n-M) \tag{5.2}$$

This recursive type of equation represents an infinite impulse response (IIR) filter. The output depends on the inputs as well as past outputs (with feedback). The output $y(n)$, at time n, depends not only on the current input $x(n)$, at time n,

135

and on past inputs $x(n-1), x(n-2), \ldots, x(n-N)$, but also on past outputs $y(n-1), y(n-2), \ldots, y(n-M)$.

If we assume all initial conditions to be zero in (5.2), the z-transform of (5.2) becomes

$$Y(z) = a_0 X(z) + a_1 z^{-1} X(z) + a_2 z^{-2} X(z) + \ldots + a_N z^{-N} X(z)$$
$$- b_1 z^{-1} Y(z) - b_2 z^{-2} Y(z) - \ldots - b_M z^{-M} Y(z) \tag{5.3}$$

Let $N = M$ in (5.3); then the transfer function $H(z)$ is

$$H(z) = \frac{Y(z)}{X(z)} = \frac{a_0 + a_1 z^{-1} + a_2 z^{-2} + \ldots + a_N z^{-N}}{1 + b_1 z^{-1} + b_2 z^{-2} + \ldots + b_N z^{-N}} = \frac{N(z)}{D(z)} \tag{5.4}$$

where $N(z)$ and $D(z)$ represent the numerator and denominator polynomial, respectively. Multiplying and dividing by z^N, $H(z)$ becomes

$$H(z) = \frac{a_0 z^N + a_1 z^{N-1} + a_2 z^{N-2} + \ldots + a_N}{z^N + b_1 z^{N-1} + b_2 z^{N-2} + \ldots + b_N} = C \prod_{i=1}^{N} \frac{z - z_i}{z - p_i} \tag{5.5}$$

which is a transfer function with N zeros and N poles. If all the coefficients b_j in (5.5) are zero, then this transfer function reduces to the transfer function with N poles at the origin in the z-plane representing the FIR filter discussed in Chapter 4. For a system to be stable, all the poles must reside inside the unit circle, as discussed in Chapter 4. Hence, for an IIR filter to be stable, the magnitude of each of its poles must be less than 1, or

a) if $|p_i| < 1$, then $h(n) \rightarrow 0$, as $n \rightarrow \infty$, yielding a stable system
b) if $|p_i| > 1$, then $h(n) \rightarrow \infty$, as $n \rightarrow \infty$, yielding an unstable system

If $|p_i| = 1$, then the system is marginally stable, yielding an oscillatory response. Furthermore, multiple-order poles on the unit circle yields an unstable system. Note again that with all the coefficients $b_j = 0$, the system reduces to a nonrecursive and stable FIR filter.

5.2 IIR FILTER STRUCTURES

There are several structures that can represent an IIR filter, as will be discussed now.

Direct Form I Structure

With the direct form I structure shown in Figure 5.1, the filter in (5.2) can be realized. There is an implied summer (not shown) in Figure 5.1. For an Nth-order

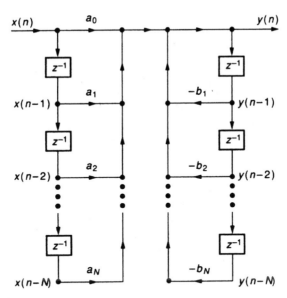

FIGURE 5.1 Direct form I IIR filter structure.

filter, this structure has $2N$ delay elements, represented by z^{-1}. For example, a second-order filter with $N = 2$ will have four delay elements.

Direct Form II Structure

The direct form II structure shown in Figure 5.2 is one of the most commonly used structures. It requires half as many delay elements as the direct form I. For example, a second-order filter requires two delay elements z^{-1}, as opposed to four with the direct form I. To show that (5.2) can be realized with the direct form II, let a delay variable $U(z)$ be defined as

$$U(z) = \frac{X(z)}{D(z)} \tag{5.6}$$

where $D(z)$ is the denominator polynomial of the transfer function in (5.4). From (5.4) and (5.6), $Y(z)$ becomes

$$Y(z) = \frac{N(z)X(z)}{D(z)} = N(z)U(z)$$

$$= U(z)\{a_0 + a_1 z^{-1} + a_2 z^{-2} + \ldots + a_N z^{-N}\} \tag{5.7}$$

where $N(z)$ is the numerator polynomial of the transfer function in (5.4). From (5.6)

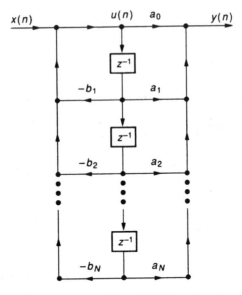

FIGURE 5.2 Direct form II IIR filter structure.

$$X(z) = U(z)D(z) = U(z)\{1 + b_1 z^{-1} + b_2 z^{-2} + \ldots + b_N z^{-N}\} \tag{5.8}$$

Taking the inverse z-transform of (5.8)

$$x(n) = u(n) + b_1 u(n-1) + b_2 u(n-2) + \ldots + b_N u(n-N) \tag{5.9}$$

Solving for $u(n)$ in (5.9)

$$u(n) = x(n) - b_1 u(n-1) - b_2 u(n-2) - \ldots - b_N u(n-N) \tag{5.10}$$

Taking the inverse z-transform of (5.7) yields

$$y(n) = a_0 u(n) + a_1 u(n-1) + a_2 u(n-2) + \ldots + a_N u(n-N) \tag{5.11}$$

The direct form II structure can be represented by (5.10) and (5.11). The delay variable $u(n)$ at the middle top of Figure 5.2 satisfies (5.10), and the output $y(n)$ in Figure 5.2 satisfies (5.11).

Equations (5.10) and (5.11) are used to program an IIR filter. Initially, $u(n-1)$, $u(n-2)$, ... are set to zero. At time n, a new sample $x(n)$ is acquired, and (5.10) is used to solve for $u(n)$. The filter's output at time n then becomes

$$y(n) = a_0 u(n) + 0$$

At time $n + 1$, a newer sample $x(n + 1)$ is acquired and the delay variables in (5.10) are updated, or

$$u(n + 1) = x(n + 1) - b_1u(n) - 0$$

where $u(n - 1)$ is updated to $u(n)$. From (5.11), the output at time $n + 1$ is

$$y(n + 1) = a_0u(n + 1) + a_1u(n) + 0$$

and so on, for time $n + 2, n + 3, \ldots$, when, for each specific time, a new input sample is acquired and the delay variables and then the output are calculated using (5.10), and (5.11), respectively.

Direct Form II Transpose

The direct form II transpose structure is a modified version of the direct form II and requires the same number of delay elements. The following steps yield a transpose structure from a direct form II version:

1. Reverse the directions of all the branches.
2. Reverse the roles of the input and output (input $\leftrightarrow$ output).
3. Redraw the structure such that the input node is on the left and the output node is on the right (as is typically done).

The direct form II transpose structure is shown in Figure 5.3. To verify this, let $u_0(n)$ and $u_1(n)$ be as shown in Figure 5.3. Then, from the transpose structure,

$$u_0(n) = a_2x(n) - b_2y(n) \tag{5.12}$$

$$u_1(n) = a_1x(n) - b_1y(n) + u_0(n - 1) \tag{5.13}$$

$$y(n) = a_0x(n) + u_1(n - 1) \tag{5.14}$$

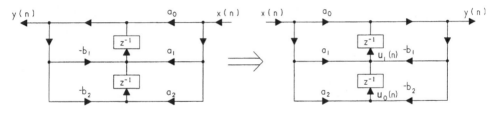

FIGURE 5.3 Direct form II transpose IIR filter structure.

Equation (5.13) becomes, using (5.12) to find $u_0(n-1)$

$$u_1(n) = a_1x(n) - b_1y(n) + [a_2x(n-1) - b_2y(n-1)] \qquad (5.15)$$

Equation (5.14) becomes, using (5.15) to solve for $u_1(n-1)$

$$y(n) = a_0x(n) + [a_1x(n-1) - b_1y(n-1) + a_2x(n-2) - b_2y(n-2)] \qquad (5.16)$$

which is the same general input-output equation (5.2) for a second-order system. This transposed structure implements first the zeros and then the poles, whereas the direct form II structure implements the poles first.

Cascade Structure

The transfer function in (5.5) can be factored as

$$H(z) = C\, H_1(z)H_2(z)\ldots H_r(z) \qquad (5.17)$$

in terms of first- or second-order transfer functions. The cascade (or series) structure is shown in Figure 5.4. An overall transfer function can be represented with cascaded transfer functions. For each section, the direct form II structure or its transpose version can be used. Figure 5.5 shows a fourth-order IIR structure in terms of two direct form II second-order sections in cascade. The transfer function $H(z)$, in terms of cascaded second-order transfer functions, can be written as

$$H(z) = \prod_{i=1}^{N/2} \frac{a_{0i} + a_{1i}z^{-1} + a_{2i}z^{-2}}{1 + b_{1i}z^{-1} + b_{2i}z^{-2}} \qquad (5.18)$$

where the constant C in (5.17) is incorporated into the coefficients, and each section is represented by i. For example, $N = 4$ for a fourth-order transfer function, and (5.18) becomes

$$H(z) = \frac{(a_{01} + a_{11}z^{-1} + a_{21}z^{-2})(a_{02} + a_{12}z^{-1} + a_{22}z^{-2})}{(1 + b_{11}z^{-1} + b_{21}z^{-2})(1 + b_{12}z^{-1} + b_{22}z^{-2})} \qquad (5.19)$$

as can be verified in Figure 5.5. From a mathematical standpoint, the proper ordering of the numerator and denominator factors does not affect the output re-

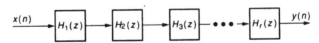

FIGURE 5.4 Cascade form IIR filter structure.

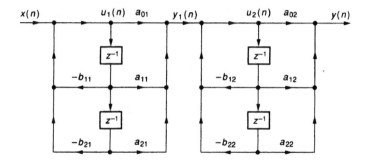

FIGURE 5.5 Fourth-order IIR filter with two direct form II sections in cascade.

sult. However, from a practical standpoint, proper ordering of each second-order section can minimize quantization noise [1–5]. Note that the output of the first section, $y_1(n)$, becomes the input to the second section. With an intermediate output result stored in one of the 40-bit wide extended precision registers, a premature truncation of the intermediate output becomes negligible. A programming example will illustrate the implementation of a sixth-order IIR filter cascaded into three second-order direct form II sections.

Parallel Form Structure

The transfer function in (5.5) can be represented as

$$H(z) = C + H_1(z) + H_2(z) + \ldots + H_r(z) \tag{5.20}$$

which can be obtained using a partial fraction expansion (PFE) on (5.5). This parallel form structure is shown in Figure 5.6. Each of the transfer functions $H_1(z), H_2(z), \ldots$ can be either first- or second-order functions. As with the cascade structure, the parallel form can be efficiently represented in terms of second-order direct form II structure sections. $H(z)$ can be expressed as

$$H(z) = C + \sum_{i=1}^{N/2} \frac{a_{0i} + a_{1i}z^{-1} + a_{2i}z^{-2}}{1 + b_{1i}z^{-1} + b_{2i}z^{-2}} \tag{5.21}$$

For example, for a fourth-order transfer function, $H(z)$ in (5.21) becomes

$$H(z) = C + \frac{a_{0i} + a_{11}z^{-1} + a_{21}z^{-2}}{1 + b_{11}z^{-1} + b_{21}z^{-2}} + \frac{a_{02} + a_{12}z^{-1} + a_{22}z^{-2}}{1 + b_{12}z^{-1} + b_{22}z^{-2}} \tag{5.22}$$

This fourth-order parallel structure is represented in terms of two direct form II sections as shown in Figure 5.7. From Figure 5.7, the output $y(n)$ can be expressed in terms of the output of each section, or

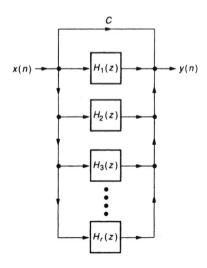

FIGURE 5.6 Parallel form IIR filter structure.

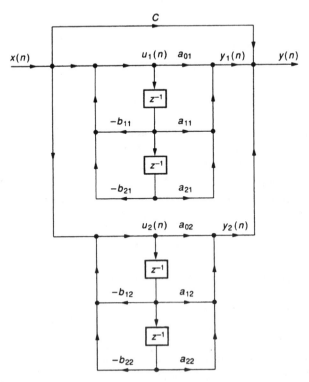

FIGURE 5.7 Fourth-order IIR filter with two direct form II sections in parallel.

$$y(n) = Cx(n) + \sum_{i=1}^{N/2} y_i(n) \tag{5.23}$$

There are other structures, such as the lattice structure, which is useful for applications in speech and adaptive filtering. Although such structure is not as computationally efficient as the direct form II or cascade structures, requiring more multiplication operations, it is less sensitive to quantization effects [6–8].

5.3 BILINEAR TRANSFORMATION

The bilinear transformation (BLT) is the most commonly used technique for transforming an analog filter into a discrete filter. It provides a one-to-one mapping from the analog s-plane to the digital z-plane, using

$$s = K\frac{z-1}{z+1} \tag{5.24}$$

The constant K in (5.24) is commonly chosen as $K = 2/T$ where T represents a sampling variable. Other values for K can be selected, since it has no consequence in the design procedure. We will choose $T = 2$, or $K = 1$ for convenience, to illustrate the bilinear transformation procedure. Solving for z in (5.24)

$$z = \frac{1+s}{1-s} \tag{5.25}$$

This transformation allows the following.

1. The left region in the s-plane, corresponding to $\sigma < 0$, maps *inside* the unit circle in the z-plane.
2. The right region in the s-plane, corresponding to $\sigma > 0$, maps *outside* the unit circle in the z-plane.
3. The imaginary $j\omega$ axis in the s-plane maps *on* the unit circle in the z-plane.

Let ω_A and ω_D represent the analog and digital frequencies, respectively. With $s = j\omega_A$ and $z = e^{j\omega_D T}$, (5.24) becomes

$$j\omega_A = \frac{e^{j\omega_D T} - 1}{e^{j\omega_D T} + 1} = \frac{e^{j\omega_D T/2}\{e^{j\omega_D T/2} - e^{-j\omega_D T/2}\}}{e^{j\omega_D T/2}\{e^{j\omega_D T/2} + e^{-j\omega_D T/2}\}} \tag{5.26}$$

Using Euler's expressions for sine and cosine in terms of complex exponential functions, ω_A from (5.26) becomes

$$\omega_A = \tan \frac{\omega_D T}{2} \tag{5.27}$$

which relates the analog frequency ω_A to the digital frequency ω_D. This relationship is plotted in Figure 5.8 for positive values of ω_A. The region corresponding to ω_A between 0 and 1 is mapped into the region corresponding to ω_D between 0 and $\omega_s/4$ in a fairly linear fashion, where ω_s is the sampling frequency in radians. However, the entire region of $\omega_A > 1$ is quite nonlinear, mapping into the region corresponding to ω_D between $\omega_s/4$ and $\omega_s/2$. This compression within this region is referred to as *frequency warping*. As a result, prewarping is done to compensate for this frequency warping. The frequencies ω_A and ω_D are such that

$$H(s)\big|_{s = j\omega_A} = H(z)\big|_{z = e^{j\omega_D T}} \tag{5.28}$$

Bilinear Transformation Design Procedure

The bilinear transformation design procedure makes use of a known analog transfer function for the design of a discrete-time filter. It can be applied using well-documented analog filter functions (Butterworth, Chebychev, etc.). Several types of filter design are available with the packages described in Appendix B. Chebyshev Type I and II provide equiripple responses in the passbands and stopbands, respectively. For a given specification, these filters have lower-order than the Butterworth-type filters, which have monotonic responses in both passbands and stopbands. An Elliptic design has equiripple in both bands, and achieve a lower-order than a Chebyshev-type design; howev-

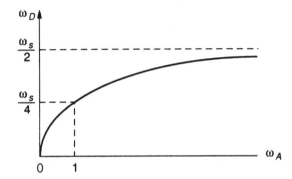

FIGURE 5.8 Relationship between analog and digital frequencies.

er, it is more difficult to design, with a highly nonlinear-phase response in the passbands. Although a Butterworth design requires a higher-order, it has a linear phase in the passbands.

Take the following steps in order to use the BLT technique and find $H(z)$.

1. Obtain a known analog transfer function $H(s)$.

2. Prewarp the desired digital frequency ω_D in order to obtain the analog frequency ω_A in (5.27).

3. Scale the frequency of the selected analog transfer function $H(s)$, using

$$H(s)|_{s=s/\omega_A} \qquad (5.29)$$

4. Obtain $H(z)$ using the BLT equation (5.24), or

$$H(z) = H(s/\omega_A)|_{s=(z-1)/(z+1)} \qquad (5.30)$$

In the case of bandpass and bandstop filters with lower and upper cutoff frequencies ω_{D1} and ω_{D2}, the two analog frequencies ω_{A1} and ω_{A2} need to be solved. The following exercises illustrate the BLT procedure.

Exercise 5.1 First-Order IIR Lowpass Filter

Given a first-order lowpass analog transfer function $H(s)$, a corresponding discrete-time filter with transfer function $H(z)$ can be obtained. Let the bandwidth or cutoff frequency $BW = 1$ r/s and the sampling frequency $F_s = 10$ Hz.

1. Choose an appropriate transfer function

$$H(s) = \frac{1}{s+1}$$

which represents a lowpass filter with a bandwidth of 1 r/s.

2. Prewarp ω_D using

$$\omega_A = \tan \frac{\omega_D T}{2} = \tan (1/20) \cong 1/20$$

where $\omega_D = B = 1$ r/s, and $T = 1/10$.

3. Scale $H(s)$ to obtain

$$H(s/\omega_A) = \frac{1}{20s+1}$$

4. Obtain the desired transfer function $H(z)$, or

$$H(z) = H(s/\omega_A)|_{s = (z-1)/(z+1)} = \frac{z+1}{21z-19}$$

Exercise 5.2 First-Order IIR Highpass Filter

Given a highpass transfer function $H(s) = s/(s+1)$, obtain a corresponding transfer function $H(z)$. Let the bandwidth or cutoff frequency be 1 r/s and the sampling frequency be 5 Hz. From the previous procedure, $H(z)$ is found to be

$$H(z) = \frac{10(z-1)}{11z-9}$$

Exercise 5.3 Second-Order IIR Bandstop Filter

Given a second-order analog transfer function $H(s)$ for a bandstop filter, a corresponding discrete-time transfer function $H(z)$ can be obtained. Let the lower and upper cutoff frequencies be 950 and 1050 Hz, respectively, with a sampling frequency F_s of 5 kHz.

The selected transfer function for a bandstop filter is

$$H(s) = \frac{s^2 + \omega_r^2}{s^2 + sB + \omega_r^2}$$

where B and ω_r are the bandwidth and center frequencies, respectively. The analog frequencies are

$$\omega_{A1} = \tan \frac{\omega_{D1}T}{2} = \tan \frac{2\pi \times 950}{2 \times 5000} = 0.6796$$

$$\omega_{A2} = \tan \frac{\omega_{D2}T}{2} = \tan \frac{2\pi \times 1050}{2 \times 5000} = 0.7756$$

The bandwidth $B = \omega_{A2} - \omega_{A1} = 0.096$, and $\omega_r^2 = (\omega_{A1})(\omega_{A2}) = 0.5271$. The transfer function $H(s)$ becomes

$$H(s) = \frac{s^2 + 0.5271}{s^2 + 0.096s + 0.5271}$$

and the corresponding transfer function $H(z)$ can be obtained with $s = (z-1)/(z+1)$, or

$$H(z) = \frac{\{(z-1)/(z+1)\}^2 + 0.5271}{\{(z-1)/(z+1)\}^2 + 0.096(z-1)/(z+1) + 0.5271}$$

which can be reduced to

$$H(z) = \frac{0.9408 - 0.5827z^{-1} + 0.9408z^{-2}}{1 - 0.5827z^{-1} + 0.8817z^{-2}} \qquad (5.31)$$

As shown later, $H(z)$ can be verified using the program BLT.BAS (on the accompanying disk), or MATLAB, which calculates $H(z)$ from $H(s)$ using the BLT technique, as we will illustrate. This can be quite useful in applying this procedure for higher-order filters.

Exercise 5.4 Fourth-Order IIR Bandpass Filter

A fourth-order IIR bandpass filter can be obtained using the BLT procedure. Let the upper and lower cutoff frequencies be 1 kHz and 1.5 kHz, respectively, and the sampling frequency be 10 kHz.

1. The transfer function $H(s)$ of a fourth-order Butterworth bandpass filter can be obtained from the transfer function of a second-order Butterworth lowpass filter, or

$$H(s) = H_{LP}(s)|_{s=(s^2+\omega_r^2)/sB}$$

where $H_{LP}(s)$ is the transfer function of a second-order Butterworth lowpass filter. $H(s)$ then becomes

$$H(s) = \frac{1}{s^2 + \sqrt{2}s + 1}\bigg|_{s=(s^2+\omega_r^2)/sB}$$

$$= \frac{s^2 B^2}{s^4 + \sqrt{2}Bs^3 + (2\omega_r^2 + B^2)s^2 + \sqrt{2}B\omega_r^2 s + \omega_r^4} \qquad (5.32)$$

2. The analog frequencies ω_{A1} and ω_{A2} are

$$\omega_{A1} = \tan\frac{\omega_{D1}T}{2} = \tan\frac{2\pi \times 1000}{2 \times 10,000} = 0.3249$$

$$\omega_{A2} = \tan\frac{\omega_{D2}T}{2} = \tan\frac{2\pi \times 1500}{2 \times 10,000} = 0.5095$$

3. The center frequency ω_r and the bandwidth B can now be found, or

$$\omega_r^2 = (\omega_{A1})(\omega_{A2}) = 0.1655$$

$$B = \omega_{A2} - \omega_{A1} = 0.1846$$

4. The analog transfer function $H(s)$ in (5.32) reduces to

$$H(s) = \frac{0.03407s^2}{s^4 + 0.26106s^3 + 0.36517s^2 + 0.04322s + 0.0274} \tag{5.33}$$

5. The corresponding $H(z)$ becomes

$$H(z) = \frac{0.02008 - 0.04016z^{-2} + 0.02008z^{-4}}{1 - 2.5495z^{-1} + 3.2021z^{-2} - 2.0359z^{-3} + 0.64137z^{-4}} \tag{5.34}$$

which is in the form of (5.4). This can be verified using the program BLT.BAS (on disk) as illustrated next.

Utility Program BLT.BAS to Find *H(z)* from *H(s)*

The utility program BLT.BAS (on the accompanying disk), written in BASIC, converts an analog transfer function $H(s)$ into an equivalent transfer function $H(z)$ using the bilinear equation $s = (z - 1)/(z + 1)$. To verify the results in (5.31) found in Exercise 5.3 for the second-order bandstop filter, run GWBASIC, then load and run BLT.BAS. The prompts and the associated data for the a and b coefficients associated with $H(s)$ are shown in Figure 5.9 (a) and the a and b coefficients associated with the transfer function $H(z)$ are shown in Figure 5.9 (b), which verifies (5.31).

Run BLT.BAS again to verify (5.34) using the data in (5.33).

```
Enter the # of numerator coefficients (30 = Max, 0 = Exit) --> 3
        Enter a(0)s^2   --> 1
        Enter a(1)s^1   --> 0
        Enter a(2)s^0   --> 0.5271

Enter the # of denominator coefficients --> 3
        Enter b(0)s^2   --> 1
        Enter b(1)s^1   --> 0.096
        Enter b(2)s^0   --> 0.5271

Are the above coefficients correct ? (y/n) y
```
(a)

```
a(0)z^-0  =   0.94085       b(0)z^-0  =   1.00000
a(1)z^-1  =  -0.58271       b(1)z^-1  =  -0.58271
a(2)z^-2  =   0.94085       b(2)z^-2  =   0.88171
```
(b)

FIGURE 5.9 Use of BLT.BAS program for bilinear transformation: (a) coefficients in s-plane; (b) coefficients in z-plane.

Utility Program AMPLIT.CPP to Find Magnitude and Phase

The utility program AMPLIT.CPP (on the accompanying disk), written in C++, can be used to plot the magnitude and phase responses of a filter for a given transfer function $H(z)$ with a maximum order of 10. Compile (using Borland's C++ compiler) and run this program. Enter the coefficients of the transfer function associated with the second-order IIR bandstop filter in Exercise 5.3 as shown in Figure 5.10 (a). Figures 5.10 (b) and (c) show the magnitude and

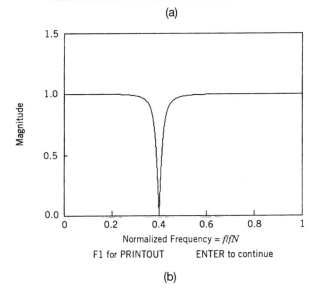

FILTER COEFFICIENTS			
NUMERATOR		DENOMINATOR	
z-0	.9408	z-0	1
z-1	-.5827	z-1	-.5827
z-2	.9408	z-2	.8817
z-3		z-3	
z-4		z-4	
z-5		z-5	
z-6		z-6	
z-7		z-7	
z-8		z-8	
z-9		z-9	
z-10		z-10	
F1 HELP	F5 Quit	F10 PLOT	

(a)

(b)

FIGURE 5.10 Use of AMPLIT.CPP program for plotting magnitude and phase: (a) coefficients in z-plane; (b) normalized magnitude; (c) normalized phase. *(Continued on next page.)*

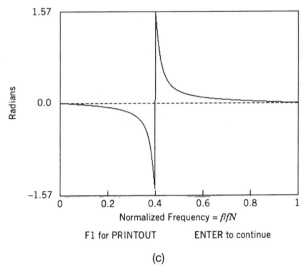

(c)

FIGURE 5.10   *(continued)*

phase of the second-order bandstop filter. From the plot of the magnitude response of $H(z)$, the normalized center frequency is shown at $v = f/F_N = 1000/2500 = 0.4$.

Run this program again to plot the magnitude response associated with the fourth-order IIR bandpass filter in Exercise 5.4. Verify the plot shown in Figure 5.11. The normalized center frequency is shown at $v = 1250/5000 = 0.25$.

A utility program MAGPHSE.BAS (on the accompanying disk), written in BASIC, can be used to tabulate the magnitude and phase responses.

5.4 PROGRAMMING EXAMPLES USING TMS320C3x AND C CODE

Several examples using both TMS320C3x and C code discuss the implementation of IIR filters. As a special case of an IIR filter with poles *on* the unit circle, a sinusoidal generation program, using the difference equation introduced in Chapter 4, is illustrated. A sixth-order IIR filter program using cascaded direct form II structures also is implemented. These programs are in both TMS320C3x and C code. Note again that the C programs can be tested/run, since the executable files are on disk, without the TMS320 floating-point assembly language tools.

Sine Generation

Using the results from Section 4.1 and Exercise 4.2 with the *z*-transform of a sinusoid, let the transfer function

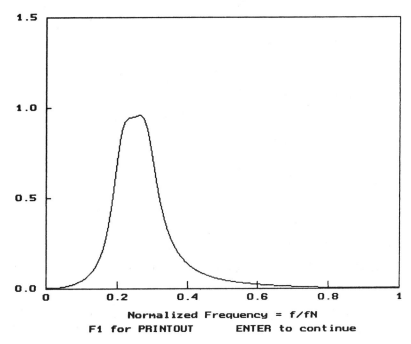

F1 for PRINTOUT **ENTER to continue**

FIGURE 5.11 Plot of magnitude response of fourth-order IIR bandpass filter using AM-PLIT.CPP.

$$H(z) = \frac{Y(z)}{X(z)} = \frac{Cz}{z^2 - Az - B} = \frac{Cz^{-1}}{1 - Az^{-1} - Bz^{-2}} \qquad (5.35)$$

where $A = 2 \cos \omega T$, $B = -1$, and $C = \sin \omega T$. Then

$$Y(z)\{1 - Az^{-1} - Bz^{-2}\} = Cz^{-1}X(z) \qquad (5.36)$$

Taking the inverse z-transform of (5.36), and assuming zero initial conditions

$$y(n) = Ay(n-1) + By(n-2) + Cx(n-1) \qquad (5.37)$$

which is a second-order recursive difference equation representing a digital oscillator. The sampling frequency or output rate of the generated sinusoidal sequence is $F_s = 1/T$, and $\omega = 2\pi f$, where f is the desired frequency of oscillation. For a given sampling frequency, we can calculate the coefficients A and C, with $B = -1$, to generate a sine function of frequency f.

If we apply an impulse at time $n = 1$, then $x(n - 1) = x(0) = 1$ for $n = 1$, and zero otherwise. With initial conditions $y(-1) = y(-2) = 0$ in (5.37)

$$n = 0: \quad y(0) = Ay(-1) + By(-2) + Cx(-1) = 0$$
$$n = 1: \quad y(1) = Ay(0) + By(-1) + Cx(0) = C$$
$$n = 2: \quad y(2) = Ay(1) + By(0) + 0 = AC$$
$$\vdots \qquad \vdots \tag{5.38}$$

For $n \geq 2$, the difference equation in (5.37) reduces to

$$y(n) = Ay(n-1) + By(n-2)$$

with $y(1) = C$ and $y(0) = 0$.

Cosine Generation

From Section 4.1, for a cosine function $\cos n\omega T$, we can let the transfer function be

$$H(z) = \frac{Y(z)}{X(z)} = \frac{z^2 - z \cos \omega T}{z^2 - 2z \cos \omega T + 1} = \frac{z^2 - (A/2)z}{z^2 - Az - B} = \frac{1 - (A/2)z^{-1}}{1 - Az^{-1} - Bz^{-2}} \tag{5.39}$$

where $A = 2 \cos \omega T$, and $B = -1$. From (5.39)

$$y(z)\{1 - Az^{-1} - Bz^{-2}\} = X(z)\{1 - (A/2)z^{-1}\} \tag{5.40}$$

Taking the inverse z-transform of (5.40), and assuming zero initial conditions

$$y(n) = Ay(n-1) + By(n-2) + x(n) - (A/2)x(n-2) \tag{5.41}$$

which represents a second-order difference equation that can be programmed to generate a cosine function. Note that the poles of this transfer function are the same as the poles associated with the transfer function for $\sin \omega T$. Hence, they are also *on* the unit circle.

Example 5.1 Sine Generation by Recursive Equation Using TMS320C3x Code

Figure 5.12 shows the program SINEA.ASM, which implements (5.37) representing a digital oscillator. The coefficients A and C are calculated and set in the program for a desired oscillation frequency of 1 kHz and a sampling or output rate of 10 kHz. The coefficient $B = -1$. For a different oscillation or sampling frequency, the coefficients A and C need to be recalculated and set. In the next example with a C code implementation, the coefficients A and C are calculated

```
;SINEA.ASM - SINE GENERATION WITH y(n)=A*y(n-1)+B*y(n-2)+C*x(n-1)
        .start   ".data",0x809900  ;starting addr of data section
        .start   ".text",0x809C00  ;starting addr of text section
        .include "AICCOM31.ASM"     ;include AIC comm routines
        .data                       ;assemble into data section
AICSEC  .word    162Ch,1h,3872h,67h    ;Fs = 10 kHz
A       .float   1.618034           ;A=2(coswT), f=1 kHz, Fs=10 kHz
B       .float   -1.0               ;B = -1
Y1      .float   0.587785           ;y(1) = C = sin(wT) = .587785
Y0      .float   0.0                ;y(0)=0
SCALER  .float   1000               ;scaling factor
        .entry   BEGIN              ;start of code
        .text                       ;assemble into text section
BEGIN   LDP      AICSEC             ;init to data page 128
        CALL     AICSET             ;initialize AIC
        LDF      @Y0,R1             ;R1=y(0)=0, out for sim
        LDF      @Y1,R1             ;R1=y(1), out for sim
        LDF      @A,R3              ;R3=A
        MPYF3    R3,R1,R1           ;R1=A*y(1), out for sim
        LDF      @Y1,R0             ;R0=y(2)
        LDF      @B,R4              ;R4=B
;y(n) for n >= 3. Output value start at n>=3
LOOP    LDF      R1,R2              ;R2=A*Y1
        MPYF3    R3,R1,R1           ;R1=A(A*Y1)
        MPYF3    R4,R0,R0           ;R0=B*Y2
        ADDF     R0,R1              ;R1=output
        LDF      R1,R5              ;save R1 for next n
        MPYF     @SCALER,R5         ;scale output amplitude
        FIX      R5,R7              ;R7=integer(R5)
        CALL     AICIO_P            ;call AIC I/O routine
        LDF      R2,R0              ;R0=A*Y1, for next n
        BR       LOOP               ;continue for each n
        .end                        ;end
```

FIGURE 5.12 Sine generation program with recursive difference equation, using TMS320C3x code (SINEA.ASM).

within the program for a specified sampling frequency and oscillation frequency. The output is scaled to obtain an appropriate output amplitude.

Although the output values $y(0)$, $y(1)$, and $y(2)$ for $n = 0$, 1, and 2 are calculated, they are not sent for output. This is acceptable in a real-time environment, with output values starting at $n \geq 3$. However, in a simulation environment, these three values contained in R1, before the looped section of code, need to be output for $n = 0$, 1, and 2. Run this program and verify an output sinusoid with a frequency of 1 kHz. The program SINESW.ASM (on disk) extends this example to yield an FM signal.

Example 5.2 Cosine Generation by Recursive Equation Using TMS320C3x Code

We can program (5.41) to generate a cosine function in a similar fashion to the previous example for the sine generation. Assume an impulse such that $x(0) = 1$, and $y(-1) = y(-2) = 0$. Then

$$n = 0: \quad y(0) = Ay(-1) + By(-2) + x(0) - (A/2)x(-1) = 1.0$$

$$n = 1: \quad y(1) = Ay(0) + By(-1) + x(1) - (A/2)x(0) = A - A/2$$

$$n = 2: \quad y(2) = Ay(1) + By(0) + 0 = A(A - A/2) + B$$

$$\vdots \qquad \vdots$$

Copy the previous program SINEA.ASM as COSINEA.ASM and edit it with the two following changes to generate a cosine function:

1. Set Y1 = 0.809017 since $y(1) = A - (A/2)$, and Y0 = $x(0)$ = 1.0
2. Add the instruction ADDF @B,R1 after the first multiplication MPYF3 instruction before the loop section of code.

Run this program and verify a cosine function of frequency 1 kHz. Note that as with the sine generation function, for simulation, the first three output samples $y(0)$, $y(1)$, and $y(2)$ contained in R1 for each n, need to be output. In this real-time implementation, the output samples are obtained for $n \geq 3$.

Example 5.3 Sine Generation by Recursive Equation Using C Code

Figure 5.13 shows the program SINEC.C, which implements (5.37) representing a digital oscillator. Note the following from the program.
   1. The sampling frequency (sample_freq) is defined or set at 10 kHz, and the desired oscillation frequency (sine_freq) is set at 3 kHz. To generate a different frequency (up to F_N), only sine_freq needs to be changed.
   2. From the main function, the coefficients A and C are calculated as follows:

$$A = 2 \cos \omega T = 2 \cos \left(\frac{2\pi \times 3000}{10{,}000} \right)$$

```
/*SINEC.C - REAL-TIME SINE GENERATION BY RECURSIVE EQUATION*/
#include "aiccomc.c"              /*AIC comm routines      */
#include "math.h"                 /*math library function */
#define sample_freq 10000         /*sample frequency       */
#define sine_freq 3000            /*desired frequency      */
#define pi 3.14159                /*constant pi            */
int AICSEC[4] = {0x162C,0x1,0x3872,0x67}; /*AIC config data*/

void sinewave(float A, float B, float C)
{
 float y[3] = {0.0,0.0,0.0};      /*y[n] array             */
 float x[3] = {0.0,0.0,1.0};      /*x[n] array             */
 int n = 2, result;              /*declare variables      */
 while(1)
   {
   TWAIT;
   y[n] = A*y[n-1] + B*y[n-2] + C*x[n-1];  /*determine y[n]*/
   result = (int)(y[n]*1000);    /*out y[n] scaled by 1000 */
   PBASE[0x48] = result << 2;    /*output to AIC           */
   y[n-2] = y[n-1];              /*shift y's back in array */
   y[n-1] = y[n];
   x[n-2] = x[n-1];              /*shift x's back in array */
   x[n-1] = x[n];
   x[n] = 0.0;                   /*set future x's to 0     */
   }
}

main()
{
 float Fs, Fosc, w, T, A, B, C;  /*declare variables       */
 AICSET();                       /*initialize AIC          */
 Fs = sample_freq;               /*get sampling frequency  */
 Fosc = sine_freq;               /*get oscillator frequency*/
 T = 1/Fs;                       /*determine sample period */
 w = 2*pi*Fosc;                  /*determine angular freq  */
 A = 2 * cos((w * T));           /*determine coefficient A */
 B = -1.0;                       /*coeff B is constant     */
 C = sin((w * T));               /*determine coefficient B */
   sinewave(A, B, C);            /*call sinewave function  */
}
```

FIGURE 5.13 Sine generation program with recursive difference equation, using C code (SINEC.C).

$$C = \sin \omega T = \sin \left(\frac{2\pi \times 3000}{10{,}000} \right)$$

3. Instead of starting at $n = 0$ and assume an impulse at $n = 1$ with $x(0) = 1$, the function `sinewave` implements the digital oscillator equation starting at $n = 2$. As such, $x(0) = x(1) = 0$, and $x(2) = 1$, as set in the x array, which produces the same result as in (5.38).

Run this program and verify a sinusoidal waveform of frequency 3 kHz. A different frequency such as 2 kHz can be readily generated by changing/setting `sine_freq` to 2000. Verify this new result if you have the floating-point assembly language tools. An AM signal can be implemented based on the program `SINEC.C`. See Experiment 5 in Section 5.5.

Example 5.4 Sixth-Order IIR Bandpass Filter Using TMS320C3x Code

This example implements a sixth-order IIR bandpass filter, centered at 1250 Hz, with a sampling frequency of 10 kHz, using a Butterworth design. The coefficients were obtained with a filter design package from Hyperception, Inc., referred to in Chapter 1. The IIR filter structure consists of three second-order direct form II stages or sections in cascade. For each stage, there are three a and two b coefficients. Figure 5.14 shows the program `IIR6BP.ASM`, which implements this filter.

The IIR function implements equations (5.10) and (5.11) obtained for the direct form II structure, with $N = 2$ for each section or stage in cascade, or

$$u(n) = x(n) - b_1 u(n-1) - b_2 u(n-2)$$

which represents the delay variable introduced in Section 5.2. The output is

$$y(n) = a_0 u(n) + a_1 u(n-1) + a_2 u(n-2)$$

where each stage output becomes the input to the subsequent stage. Run this program and verify the frequency response of the sixth-order IIR filter plotted in Figure 5.15. The frequency response of a fourth-order IIR filter is also plotted to illustrate the sharper characteristics of an IIR filter design with a higher order. This plot is obtained with a Hewlett Packard (HP) signal analyzer. Note the following.

1. For each of the three stages, the coefficients are ordered as b_1, b_2, a_1, a_2, and a_0, which correspond to $b[i][0]$, $b[i][1]$, $a[i][1]$, $a[i][2]$, and $a[i][0]$ in the program.

2. The block of code between the repeat block instruction RPTB LOOP and

```
*IIR6BP.ASM - SIXTH-ORDER IIR BANDPASS, Fc = 1250 Hz
        .start   ".text", 0x809900 ;starting address of text
        .start   ".data", 0x809C00 ;starting address of data
        .include "AICCOM31.ASM"    ;include AIC comm routines
        .entry   BEGIN             ;start of code
        .text                      ;assemble into text
BEGIN   LDP      @COEFF_ADDR        ;init to data page 128
        CALL     AICSET             ;initialize AIC
IIR     LDI      @COEFF_ADDR,AR0    ;AR0 points to coefficients address
        LDI      @DLY_ADDR,AR1      ;AR1 points to addr of delay samples
        CALL     AICIO_P            ;call AIC for polling
        FLOAT    R6,R3              ;stage input
        MPYF3    *AR0++,*AR1++,R0   ;b[i][0]*dly[i][0]
        LDI      STAGES-1, RC       ;initialize stage counter
        RPTB     LOOP               ;repeat LOOP RC times
        MPYF3    *AR0++,*AR1-,R1    ;b[i][1]*dly[i][1]
||      SUBF3    R0,R3,R3           ;input-b[i][0]*dly[i][0]
        MPYF3    *AR0++,*AR1++,R0   ;a[i][1]*dly[i][0]
||      SUBF3    R1,R3,R2 ;dly=input-b[i][0]*dly[i][0]-b[i][1]*dly[i][1]
        MPYF3    *AR0++,*AR1-,R1    ;a[i][2]*dly[i][1]
        ADDF3    R0,R1,R3           ;a[i][2]*dly[i][1]+a[i][1]*dly[i][0]
        LDF      *AR1,R4            ;dly[i][2]
||      STF      R2,*AR1++          ;dly[i][0] = dly
        MPYF3    R2,*AR0++,R2       ;dly*a[i][0]
||      STF      R4,*AR1++          ;dly[i][1] = dly[i][0]
LOOP    MPYF3    *AR0++,*AR1++,R0   ;b[i+1][0]*dly[i+1][0]
||      ADDF3    R2,R3,R3           ;stage output=input of next stage
        FIX      R3,R7              ;convert output to integer
        BR       IIR
        .data ;b[i][0]    b[i][1]    a[i][1]     a[i][2]     a[i][0]
COEFF   .float  -1.4435E+0, 9.4880E-1, 0.0000E+0, -5.3324E-2, 5.3324E-2
        .float  -1.3427E+0, 8.9515E-1, 0.0000E+0, -5.3324E-2, 5.3324E-2
        .float  -1.3082E+0, 9.4378E-1, 0.0000E+0, -5.3324E-2, 5.3324E-2
DLY        .float  0, 0, 0, 0, 0, 0  ;init delay var for each stage
STAGES     .set    3                 ;number of stages
COEFF_ADDR .word   COEFF             ;address of COEFF
DLY_ADDR   .word   DLY               ;address of DELAY
AICSEC     .word   162ch,1h,3872h,67h ;AIC config data, Fs = 10 kHz
           .end                      ;end
```

FIGURE 5.14 IIR filter program for sixth-order bandpass filter (IIR6BP.ASM).

the instruction specified by the address LOOP (which includes the parallel instruction ADDF3 R2, R3, R3) is executed three times (once for each stage).

3. After each output sample, execution branches back to the function IIR. Each input sample is acquired through R3.

4. AR0 points to the starting address of a table containing three sets of coefficients; a set for each stage, ordered as b_1, b_2, a_1, a_2, a_0. AR1 points to the starting address of another table containing the delay variables for each stage, and ordered as $u(n-1), u(n-2), \ldots$. Note that these delay variables, two for each stage, are initialized to zero.

5. Consider the following multiply operations:

  a) The first one calculates

$$R0 = b_1 u(n-1)$$

  Then, AR0 and AR1 are postincremented to point at b_2 and $u(n-2)$, respectively.

  b) The second one calculates

$$R1 = b_2 u(n-2)$$

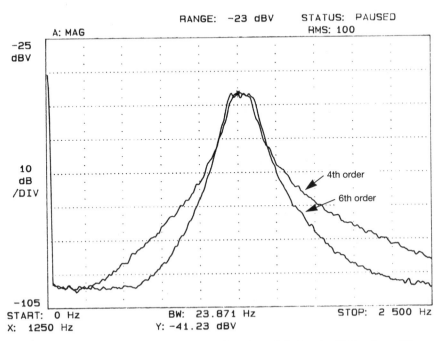

FIGURE 5.15 Plot of frequency responses of fourth- and sixth-order IIR bandpass filters.

and the subtract instruction in parallel calculates

$$R3 = x(n) - b_1u(n - 1)$$

AR0 is then incremented to point at a_1 while AR1 is decremented to point "back" at $u(n - 1)$.

c) The third one calculates

$$R0 = a_1u(n - 1)$$

The SUBF3 instruction in parallel calculates

$$R2 = x(n) - b_1u(n - 1) - b_2u(n - 2)$$

d) The fourth one calculates

$$R1 = a_2u(n - 2)$$

and the subsequent ADDF3 instruction yields

$$R3 = a_1u(n - 1) + a_2u(n - 2)$$

The LDF and STF instructions in parallel update the delay variable $u(n - 1)$ to $u(n)$.

e) The fifth one calculates

$$R2 = a_0u(n)$$

and the STF instruction in parallel updates $u(n - 2)$ to $u(n - 1)$.

f) The sixth and last one calculates

$$R0 = b_1u(n - 1)$$

for the *subsequent* stage, and the instruction ADDF3   R2,R3,R3 in parallel yields the stage output as

$$R3 = a_0u(n) + R3 \text{ (obtained from step d)}$$

This effectively implements the equation

$$y(n) = a_0u(n) + a_1u(n - 1) + a_2u(n - 2)$$

After an output sample is obtained at the last stage, AR0 and AR1 are reinitialized to point at the beginning addresses of the coefficients and delay samples, respectively, as in step 4.

Example 5.5 Sixth-Order IIR Bandpass Filter Using C Code

The program IIR6BPC.C shown in Figure 5.16 implements the same sixth-order IIR bandpass filter discussed in the previous example, using three cascaded direct form II second-order sections. The *a* and *b* coefficients for each stage are contained in the coefficient file IIR6COEF.H shown in Figure 5.17, which is "included" in the program IIR6BPC.C. The three sets of *a* and *b* coefficients (a set for each stage) are ordered as a_0, a_1, a_2, b_1, b_2. Note the following from the IIR function:

1. It is interrupt-driven, with a sampling frequency set to 10 kHz.
2. The For loop is executed three times, once for each stage, represented with the variable *i*.
3. *yn* calculates the stage output by implementing (5.11) in reverse order, or

$$y(n) = a_2u(n-2) + a_1u(n-1) + a_0u(n)$$

Then, the delay variables are updated so that $u(n-2)$ becomes $u(n-1)$, and $u(n-1)$ becomes $u(n)$. The stage output then becomes the input to the subsequent stage.

The overall filter's output is the output at the third stage. Run this program and verify identical results as in the previous implementation using TMS320C3x code.

5.5   EXPERIMENT 5: IIR FILTER DESIGN AND IMPLEMENTATION

1. Implement the cosine generation Example 5.2.
2. Three sets of coefficients associated with a fourth, a sixth, and an eighth-order IIR filter were obtained using the DigiFilter package described in Appendix B. The center and sampling frequencies for each filter are 1250 and 10,000 Hz, respectively. The filter design uses cascaded direct form II sections.

Test these coefficients using an IIR filter program such as IIR6BP.ASM, which implements a sixth-order IIR filter with three stages or sections in cascade and needs to be slightly modified for each of the three filters. Set the number of stages and initialize appropriately the delay variables through DLY. The ordering of the *a* and *b* coefficients in the program correspond to the *a* and *b* coefficients associated with the transfer function $H(z)$ in (5.4). Many authors re-

```
/*IIR6BPC.C -REAL-TIME 6th-ORDER IIR BANDPASS FILTER*/
#include "aiccomc.c"      /*include AIC comm routines*/
#include "iir6coef.h"     /*coefficients file        */
float dly[stages][2] = {0};   /*delay samples        */
int AICSEC[4] = {0x162C,0x1,0x3872,0x67}; /*AIC data*/
int data_in, data_out;


float IIR(int *IO_in, int *IO_out, int n, int len)
{
 int i, loop = 0;
 float un, yn, input;
 while (loop < len)
 {
  asm("     IDLE   ");
  ++loop;
  input = *IO_in;
  for (i = 0; i < n; i++)
   {
   un = input - b[i][0] * dly[i][0] - b[i][1] * dly[i][1];
   yn = a[i][2]*dly[i][1] + a[i][1]*dly[i][0] + a[i][0]*un;
   dly[i][1] = dly[i][0];
   dly[i][0] = un;
   input = yn;
   }
     *IO_out = yn;
 }
}


void c_int05()
{
  PBASE[0x48] = data_out << 2;
  data_in = PBASE[0x4C] << 16 >> 18;
}


main()
{
  #define length 345
  int *IO_OUTPUT, *IO_INPUT;
  IO_INPUT = &data_in;
  IO_OUTPUT = &data_out;
  AICSET_I();
  for (;;)
   IIR((int *)IO_INPUT, (int *)IO_OUTPUT, stages, length);
}
```

FIGURE 5.16 IIR filter program for sixth-order bandpass filter with coefficient file included (IIR6BPC.C).

```
/*IIR6COEF.H-COEFF FILE FOR SIXTH-ORDER IIR BANDPASS FILTER*/
#define stages 3                    /*number of 2nd-order stages*/
float a[stages][3]=    {      /*numerator coefficients    */
{5.3324E-02, 0.0000E+00, -5.3324E-02},     /*a10, a11, a12 */
{5.3324E-02, 0.0000E+00, -5.3324E-02},     /*a20, a21, a22 */
{5.3324E-02, 0.0000E+00, -5.3324E-02} };  /*a30, a31, a32 */
float b[stages][2]=    {     /*denominator coefficients */
{-1.4435E+00, 9.4879E-01},     /*b11, b12             */
{-1.3427E+00, 8.9514E-01},     /*b21, b22             */
{-1.3082E+00, 9.4377E-01} };   /*b31, b32             */
```

FIGURE 5.17 Coefficient file for sixth-order IIR filter program (IIR6COEF.H).

verse the a and b notation in (5.1). Verify that the eighth-order IIR filter is more selective (sharper).

a) Fourth-order Elliptic

	First stage	Second stage
a_0	0.078371	0.143733
a_1	−0.148948	0.010366
a_2	0.078371	0.143733
b_1	−1.549070	−1.228110
b_2	0.968755	0.960698

b) Sixth-order Butterworth

	First stage	Second stage	Third stage
a_0	0.137056	0.122159	0.122254
a_1	0.0	0.0	0.0
a_2	−0.137056	−0.122159	−0.122254
b_1	−1.490630	−1.152990	−1.256790
b_2	0.886387	0.856946	0.755492

c) Eighth-order Butterworth

	First stage	Second stage	Third stage	Fourth stage
a_0	0.123118	0.130612	0.127179	0.143859
a_1	0.0	0.0	0.0	0.0
a_2	−0.123118	−0.130612	−0.127179	−0.143859
b_1	−1.18334	−1.33850	−1.15014	−1.52176
b_2	0.754301	0.777976	0.884409	0.910547

3. The program SINECMOD.C (on disk) extends the sine generator program SINEC.C to implement an AM signal. Verify that the resulting spectrum contains the sum and difference of the carrier and modulation frequencies. Examine the effects of the modulation index on the carrier and the sidebands.

REFERENCES

1. L. B. Jackson, *Digital Filters and Signal Processing,* Kluwer Academic, Norwell, MA, 1996.

2. L. B. Jackson, "Roundoff Noise Analysis for Fixed-Point Digital Filters Realized in Cascade or Parallel Form," *IEEE Trans. on Audio and Electroacoustics,* Au-18, 107–122, June (1970).

3. L. B. Jackson, "An Analysis of Limit Cycles due to Multiplicative Rounding in Recursive Digital Filters," in *Proceedings of the 7th Allerton Conference on Circuit and System Theory,* 1969, 69–78.

4. L. B. Lawrence and K. V. Mirna, "A New and Interesting Class of Limit Cycles in Recursive Digital Filters," in *Proceedings of the IEEE International Symposium on Circuit and Systems,* April 1977, 191–194.

5. R. Chassaing and D. W. Horning, *Digital Signal Processing with the TMS320C25,* Wiley, New York, 1990.

6. A. H. Gray and J. D. Markel, "Digital Lattice and Ladder Filter Synthesis," in *IEEE Trans. on Acoustics, Speech, and Signal Processing,* ASSP-21, 491–500, (1973).

7. A. H. Gray and J. D. Markel, "A Normalized Digital Filter Structure," in *IEEE Trans. on Acoustics, Speech, and Signal Processing,* ASSP-23, 268–277 (1975).

8. R. Chassaing, *Digital Signal Processing with C and the TMS320C30,* Wiley, New York, 1992.

9. A. V. Oppenheim and R. Schafer, *Discrete-Time Signal Processing,* Prentice-Hall, Englewood Cliffs, NJ, 1989.

10. E. C. Ifeachor and B. W. Jervis, Digital Signal Processing A Practical Approach, Addison-Wesley, 1993.

11. N. Ahmed and T. Natarajan, Discrete-Time Signals and Systems, Reston, Reston, VA, 1983.

12. D. W. Horning and R. Chassaing, "IIR Filter Scaling for Real-Time Digital Signal Processing", in *IEEE Trans. on Education,* Feb. 1991.

13. P. A. Lynn and W. Fuerst, Introductory Digital Signal Processing With Computer Applications, Wiley, New York, 1994.

14. L. C. Ludemen, Fundamentals of Digital Signal Processing, Harper & Row, New York, 1986.

15. M. G. Bellanger, *Digital Filters and Signal Analysis,* Prentice-Hall, Englewood Cliffs, NJ, 1986.

16. F. J. Taylor, *Principles of Signals and Systems,* McGraw-Hill, New York, 1994.

17. F. J. Taylor, *Digital Filter Design Handbook,* Marcel Dekker, New York, 1983.

18. W. D. Stanley, G. R. Dougherty, and R. Dougherty, *Digital Signal Processing,* Reston, Reston, VA, 1984.

19. S. D. Stearns and R. A. David, Signal Processing in Fortran and C, Prentice-Hall, Englewood Cliffs, NJ, 1993.

20. R. Kuc, *Introduction to Digital Signal Processing,* McGraw-Hill, New York, 1988.

21. H. Baher, *Analog and Digital Signal Processing,* Wiley, New York, 1990.

22. R. A. Roberts and C. T. Mullis, *Digital Signal Processing,* Addison-Wesley, Reading, MA, 1987.

23. J. R. Johnson, *Introduction to Digital Signal Processing,* Prentice-Hall, Englewood Cliffs, NJ, 1989.

24. S. Haykin, *Modern Filters,* Macmillan, New York, 1989.

25. T. Young, *Linear Systems and Digital Signal Processing,* Prentice-Hall, Englewood Cliffs, NJ, 1985.

26. A. Ambardar, *Analog and Digital Signal Processing,* PWS, MA, 1995.

27. E. P. Cunningham, *Digital Filtering: An Introduction,* Houghton Mifflin, MA, 1992.

28. A.W.M. van den Enden and N.A.M. Verhoeckx, *Discrete-Time Signal Processing,* Prentice-Hall International (UK) Ltd, Hertfordshire, 1989.

29. M. Bellanger, *Digital Processing of Signals Theory and Practice,* Wiley, New York, 1989.

6

Fast Fourier Transform

- The fast Fourier transform using radix-2 and radix-4
- Decimation or decomposition in frequency and in time
- Programming examples

The fast Fourier transform (FFT) is an efficient algorithm that is used for converting a time-domain signal into an equivalent frequency-domain signal, based on the discrete Fourier transform (DFT). A real-time programming example is included with a main C program that calls an FFT assembly function.

6.1 INTRODUCTION

The discrete Fourier transform converts a time-domain sequence into an equivalent frequency-domain sequence. The inverse discrete Fourier transform performs the reverse operation and converts a frequency-domain sequence into an equivalent time-domain sequence. The fast Fourier transform (FFT) is a very efficient algorithm technique based on the discrete Fourier transform, but with fewer computations required. The FFT is one of the most commonly used operations in digital signal processing to provide a frequency spectrum analysis [1–6]. Two different procedures are introduced to compute an FFT: the decimation-in-frequency and the decimation-in-time. Several variants of the FFT have been used, such as the Winograd transform [7,8], the discrete cosine transform (DCT) [9], and the discrete Hartley transform [10–12]. Programs based on the DCT, FHT, and the FFT are available in [9].

6.2 DEVELOPMENT OF THE FFT ALGORITHM WITH RADIX-2

The FFT reduces considerably the computational requirements of the discrete Fourier transform (DFT). The DFT of a discrete-time signal $x(nT)$ is

$$X(k) = \sum_{n=0}^{N-1} x(n) \, W^{nk} \qquad k = 0, 1, \ldots, N-1 \tag{6.1}$$

where the sampling period T is implied in $x(n)$ and N is the frame length. The constants W are referred to as twiddle constants or factors, which represent the phase, or

$$W = e^{-j2\pi/N} \tag{6.2}$$

and is a function of the length N. Equation (6.1) can be written for $k = 0, 1, \ldots, N-1$, as

$$X(k) = x(0) + x(1)W^k + x(2)W^{2k} + \ldots + x(N-1)W^{(N-1)k} \tag{6.3}$$

This represents a matrix of $N \times N$ terms, since $X(k)$ needs to be calculated for N values of k. Since (6.3) is an equation in terms of a complex exponential, for each specific k there are approximately N complex additions and N complex multiplications. Hence, the computational requirements of the DFT can be very intensive, especially for large values of N.

The FFT algorithm takes advantage of the periodicity and symmetry of the twiddle constants to reduce the computational requirements of the FFT. From the periodicity of W

$$W^{k+N} = W^k \tag{6.4}$$

and, from the symmetry of W

$$W^{k+N/2} = -W^k \tag{6.5}$$

Figure 6.1 illustrates the properties of the twiddle constants W for $N = 8$. For example, let $k = 2$, and note that from (6.4), $W^{10} = W^2$, and from (6.5), $W^6 = -W^2$.

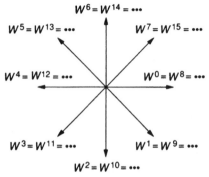

FIGURE 6.1 Periodicity and symmetry of twiddle constant W.

For a radix-2 (base 2), the FFT decomposes an N-point DFT into two $(N/2)$-point or smaller DFT's. Each $(N/2)$-point DFT is further decomposed into two $(N/4)$-point DFT's, and so on. The last decomposition consists of $(N/2)$ two-point DFT's. The smallest transform is determined by the radix of the FFT. For a radix-2 FFT, N must be a power or base of two, and the smallest transform or the last decomposition is the two-point DFT. For a radix-4, the last decomposition is a four-point DFT.

6.3 DECIMATION-IN-FREQUENCY FFT ALGORITHM WITH RADIX-2

Let a time-domain input sequence $x(n)$ be separated into two halves:

a)
$$x(0), x(1), \ldots, x\left(\frac{N}{2} - 1\right)$$
(6.6)

and

b)
$$x\left(\frac{N}{2}\right), x\left(\frac{N}{2} + 1\right), \ldots, x(N - 1)$$
(6.7)

Taking the DFT of each set of the sequence in (6.6) and (6.7),

$$X(k) = \sum_{n=0}^{(N/2)-1} x(n)W^{nk} + \sum_{n=N/2}^{N-1} x(n)W^{nk}$$
(6.8)

Let $n = n + N/2$ in the second summation of (6.8), $X(k)$ becomes

$$X(k) = \sum_{n=0}^{(N/2)-1} x(n)W^{nk} + W^{kN/2}\sum_{n=0}^{(N/2)-1} x\left(n + \frac{N}{2}\right)W^{nk}$$
(6.9)

where $W^{kN/2}$ is taken out of the second summation because it is not a function of n. Using,

$$W^{kN/2} = e^{-jk\pi} = (e^{-j\pi})^k = (\cos \pi - j\sin \pi)^k = (-1)^k$$

in (6.9), $X(k)$ becomes

$$X(k) = \sum_{n=0}^{(N/2)-1} \left[x(n) + (-1)^k x\left(n + \frac{N}{2}\right)\right]W^{nk}$$
(6.10)

Because $(-1)^k = 1$ for even k and -1 for odd k, (6.10) can be separated for even and odd k, or

for even k:
$$X(k) = \sum_{n=0}^{(N/2)-1} \left[x(n) + x\left(n + \frac{N}{2}\right) \right] W^{nk} \tag{6.11}$$

for odd k:
$$X(k) = \sum_{n=0}^{(N/2)-1} \left[x(n) - x\left(n + \frac{N}{2}\right) \right] W^{nk} \tag{6.12}$$

Substituting $k = 2k$ for even k, and $k = 2k + 1$ for odd k, (6.11) and (6.12) can be written as, for $k = 0, 1, \ldots, (N/2) - 1$,

$$X(2k) = \sum_{n=0}^{(N/2)-1} \left[x(n) + x\left(n + \frac{N}{2}\right) \right] W^{2nk} \tag{6.13}$$

$$x(2K + 1) = \sum_{n=0}^{(N/2)-1} \left[x(n) - x\left(n + \frac{N}{2}\right) \right] W^n W^{2nk} \tag{6.14}$$

Because the twiddle constant W is a function of the length N, it can be represented as W_N. Then, W_N^2 can be written as $W_{N/2}$. Let

$$a(n) = x(n) + x(n + N/2) \tag{6.15}$$

$$b(n) = x(n) - x(n + N/2) \tag{6.16}$$

Equations (6.13) and (6.14) can be more clearly written as two $(N/2)$-point DFT's, or

$$X(2k) = \sum_{n=0}^{(N/2)-1} a(n) W_{N/2}^{nk} \tag{6.17}$$

$$X(2k + 1) = \sum_{n=0}^{(N/2)-1} b(n) W_N^n W_{N/2}^{nk} \tag{6.18}$$

Figure 6.2 shows the decomposition of an N-point DFT into two $(N/2)$-point DFT's, for $N = 8$. As a result of the decomposition process, the X's in Figure 6.2 are even in the upper half and they are odd in the lower half. The decomposition process can now be repeated such that each of the $(N/2)$-point DFT's is further decomposed into two $(N/4)$-point DFT's, as shown in Figure 6.3, again using $N = 8$ to illustrate.

The upper section of the output sequence in Figure 6.2 yields the sequence $X(0)$ and $X(4)$ in Figure 6.3, ordered as even. $X(2)$ and $X(6)$ from Figure 6.3 represent the odd values. Similarly, the lower section of the output sequence in Figure 6.2 yields $X(1)$ and $X(5)$, ordered as the even values, and $X(3)$ and $X(7)$ as the odd values. This scrambling is due to the decomposition process. The final

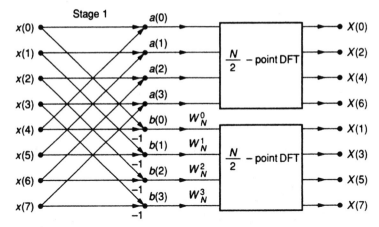

FIGURE 6.2 Decomposition of N-point DFT into two ($N/2$)-point DFT's, for $N = 8$.

order of the output sequence $X(0)$, $X(4)$, ... in Figure 6.3 is shown to be scrambled. The output needs to be resequenced or reordered. A special instruction using indirect addressing with bit-reversal, introduced in Chapter 2 in conjunction with circular buffering, is available on the TMS320C3x to reorder such a sequence. The output sequence $X(k)$ represents the DFT of the time sequence $x(n)$.

This is the last decomposition, since we have now a set of ($N/2$) two-point DFT's, the lowest decomposition for a radix-2. For the two-point DFT, $X(k)$ in (6.1) can be written as

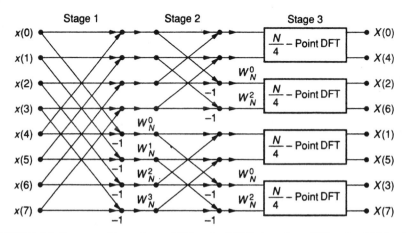

FIGURE 6.3 Decomposition of two ($N/2$)-point DFT's into four ($N/4$)-point DFT's, for $N = 8$.

$$X(k) = \sum_{n=0}^{1} x(n)W^{nk} \qquad k = 0, 1 \qquad (6.19)$$

or

$$X(0) = x(0)W^0 + x(1)W^0 = x(0) + x(1) \qquad (6.20)$$

$$X(1) = x(0)W^0 + x(1)W^1 = x(0) - x(1) \qquad (6.21)$$

since $W^1 = e^{-j2\pi/2} = -1$. Equations (6.20) and (6.21) can be represented by the flow graph in Figure 6.4, usually referred to as a butterfly. The final flow graph of an eight-point FFT algorithm is shown in Figure 6.5. This algorithm is referred as decimation-in-frequency (DIF) because the output sequence $X(k)$ is decomposed (decimated) into smaller subsequences, and this process continues through M stages or iterations, where $N = 2^M$. The output $X(k)$ is complex with both real and imaginary components, and the FFT algorithm can accomodate either complex or real input values.

The FFT is not an approximation of the DFT. It yields the same result as the DFT with less computations required. This reduction becomes more and more important with higher-order FFT.

There are other FFT structures that have been used to illustrate the FFT. An alternative flow graph to the one shown in Figure 6.5 can be obtained with ordered output and scrambled input.

An eight-point FFT is illustrated through an exercise as well as through a programming example. We will see that flow graphs for higher-order FFT (larger N) can readily be obtained.

Exercise 6.1 Eight-Point FFT Using Decimation-in-Frequency

Let the input $x(n)$ represent a rectangular waveform, or $x(0) = x(1) = x(2) = x(3) = 1$, and $x(4) = x(5) = x(6) = x(7) = 0$. The eight-point FFT flow graph in Figure 6.5 can be used to find the output sequence $X(k)$, $k = 0, 1, \ldots, 7$. With $N = 8$, four twiddle constants need to be calculated, or

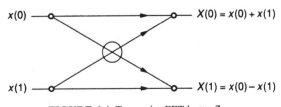

FIGURE 6.4 Two-point FFT butterfly.

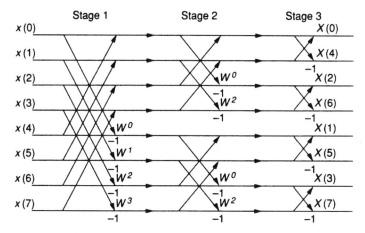

FIGURE 6.5 Eight-point FFT flow graph using decimation-in-frequency.

$$W^0 = 1$$

$$W^1 = e^{-j2\pi/8} = \cos(\pi/4) - j\sin(\pi/4) = 0.707 - j0.707$$

$$W^2 = e^{-j4\pi/8} = -j$$

$$W^3 = e^{-j6\pi/8} = -0.707 - j0.707$$

The intermediate output sequence can be found after each stage.

1. At stage 1:

$$x(0) + x(4) = 1 \rightarrow x'(0)$$

$$x(1) + x(5) = 1 \rightarrow x'(1)$$

$$x(2) + x(6) = 1 \rightarrow x'(2)$$

$$x(3) + x(7) = 1 \rightarrow x'(3)$$

$$[x(0) - x(4)]W^0 = 1 \rightarrow x'(4)$$

$$[x(1) - x(5)]W^1 = 0.707 - j0.707 \rightarrow x'(5)$$

$$[x(2) - x(6)]W^2 = -j \rightarrow x'(6)$$

$$[x(3) - x(7)]W^3 = -0.707 - j0.707 \rightarrow x'(7)$$

where $x'(0), x'(1), \ldots, x'(7)$ represent the intermediate output sequence after the first iteration that becomes the input to the second stage.

2. At stage 2:

$$x'(0) + x'(2) = 2 \rightarrow x''(0)$$
$$x'(1) + x'(3) = 2 \rightarrow x''(1)$$
$$[x'(0) - x'(2)]W^0 = 0 \rightarrow x''(2)$$
$$[x'(1) - x'(3)]W^2 = 0 \rightarrow x''(3)$$
$$x'(4) + x'(6) = 1 - j \rightarrow x''(4)$$
$$x'(5) + x'(7) = (0.707 - j0.707) + (-0.707 - j0.707) = -j1.41 \rightarrow x''(5)$$
$$[x'(4) - x'(6)]W^0 = 1 + j \rightarrow x''(6)$$
$$[x'(5) - x'(7)]W^2 = -j1.41 \rightarrow x''(7)$$

The resulting intermediate, second-stage output sequence $x''(0)$, $x''(1)$, ..., $x''(7)$ becomes the input sequence to the third stage.

3. **At stage 3:**

$$X(0) = x''(0) + x''(1) = 4$$
$$X(4) = x''(0) - x''(1) = 0$$
$$X(2) = x''(2) + x''(3) = 0$$
$$X(6) = x''(2) - x''(3) = 0$$
$$X(1) = x''(4) + x''(5) = (1 - j) + (-j1.41) = 1 - j2.41$$
$$X(5) = x''(4) - x''(5) = 1 + j0.41$$
$$X(3) = x''(6) + x''(7) = (1 + j) + (-j1.41) = 1 - j0.41$$
$$X(7) = x''(6) - x''(7) = 1 + j2.41$$

We now use the notation of X's to represent the final output sequence. The values $X(0)$, $X(1)$, ..., $X(7)$ form the scrambled output sequence. These results can be verified with an FFT function available with the MATLAB software package described in Appendix B. We will show soon how to reorder the output sequence and plot the output magnitude.

Exercise 6.2 Sixteen-Point FFT

Given $x(0) = x(1) = \ldots = x(7) = 1$, and $x(8) = x(9) = \ldots = x(15) = 0$, which represents a rectangular input sequence. The output sequence can be found using the 16-point flow graph shown in Figure 6.6. The intermediate output results after each stage are found in a similar manner to the previous example. Eight twiddle constants W^0, W^1, ..., W^7 need to be calculated for $N = 16$.

Verify the scrambled output sequence X's as shown in Figure 6.6. Reorder this output sequence and take its magnitude. Verify the plot in Figure 6.7, which

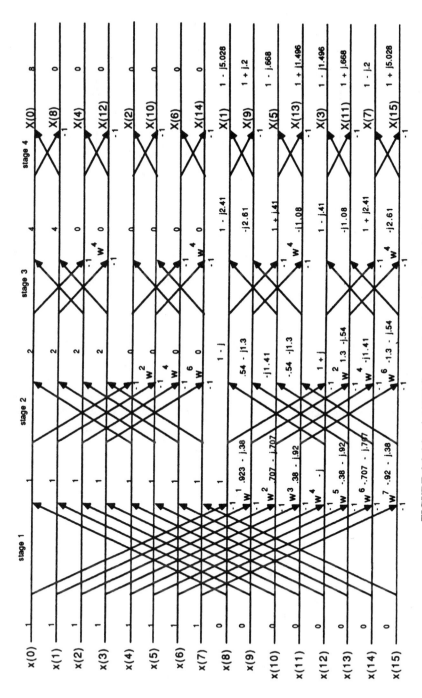

FIGURE 6.6 16-point FFT flow graph using decimation-in-frequency.

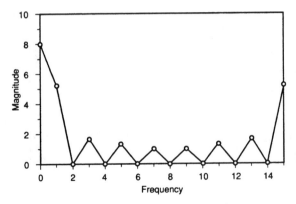

FIGURE 6.7 Output magnitude for 16-point FFT.

represents a sinc function. The output $X(8)$ represents the magnitude at the Nyquist frequency. These results can be verified with an FFT function available with MATLAB, described in Appendix B.

6.4 DECIMATION-IN-TIME FFT ALGORITHM WITH RADIX-2

Whereas the decimation-in-frequency (DIF) process decomposes an output sequence into smaller subsequences, the decimation-in-time (DIT) is another process that decomposes the input sequence into smaller subsequences. Let the input sequence be decomposed into an even sequence and an odd sequence, or

$$x(0), x(2), x(4), \ldots, x(2n)$$

and

$$x(1), x(3), x(5), \ldots, x(2n+1)$$

We can apply (6.1) to these two sequences to obtain

$$X(k) = \sum_{n=0}^{(N/2)-1} x(2n)W^{2nk} + \sum_{n=0}^{(N/2)-1} x(2n+1)W^{(2n+1)k} \qquad (6.22)$$

Using $W_N^2 = W_{N/2}$ in (6.22)

$$X(k) = \sum_{n=0}^{(N/2)-1} x(2n)W_{N/2}^{nk} + W_N^k \sum_{n=0}^{(N/2)-1} x(2n+1)W_{N/2}^{nk} \qquad (6.23)$$

which represents two $(N/2)$-point DFT's. Let

$$C(k) = \sum_{n=0}^{(N/2)-1} x(2n)W_{N/2}^{nk} \qquad (6.24)$$

$$D(k) = \sum_{n=0}^{(N/2)-1} X(2n+1)W_{N/2}^{nk} \qquad (6.25)$$

Then $X(k)$ in (6.23) can be written as

$$X(k) = C(k) + W_N^k D(k) \qquad (6.26)$$

Equation (6.26) needs to be interpreted for $k > (N/2) - 1$. Using the symmetry property (6.5) of the twiddle constant, $W^{k+N/2} = -W^k$,

$$X(k + N/2) = C(k) - W^k D(k) \qquad k = 0, 1, \ldots, (N/2) - 1 \qquad (6.27)$$

For example, for $N = 8$, (6.26) and (6.27) become

$$X(k) = C(k) + W^k D(k) \qquad k = 0, 1, 2, 3 \qquad (6.28)$$

$$X(k + 4) = C(k) - W^k D(k) \qquad k = 0, 1, 2, 3 \qquad (6.29)$$

Figure 6.8 shows the decomposition of an eight-point DFT into two four-point DFT's with the decimation-in-time procedure. This decomposition or decimation process is repeated so that each four-point DFT is further decomposed into

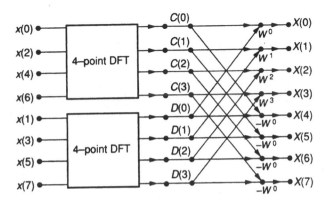

FIGURE 6.8 Decomposition of eight-point DFT into two four-point DFT's using DIT.

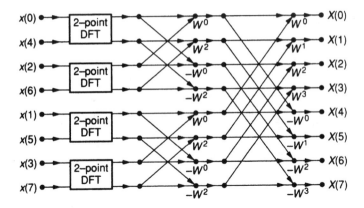

FIGURE 6.9 Decomposition of two four-point DFT's into four two-point DFT's using DIT.

two two-point DFT's, as shown in Figure 6.9. Since the last decomposition is $(N/2)$ two-point DFTs, this is as far as this process goes.

Figure 6.10 shows the final flow graph for an eight-point FFT using a decimation-in-time process. The input sequence is shown to be scrambled in Figure 6.10, in the same manner as the output sequence $X(k)$ was scrambled during the decimation-in-frequency process. With the input sequence $x(n)$ scrambled, the resulting output sequence $X(k)$ becomes properly ordered. Identical results are obtained with an FFT using either the decimation-in-frequency (DIF) or the decimation-in-time (DIT) process.

An alternative DIT flow graph to the one shown in Figure 6.10, with ordered input and scrambled output, also can be obtained.

The following exercise shows that the same results are obtained for an eight-point FFT with the DIT process as in Exercise 6.1 with the DIF process.

Exercise 6.3 Eight-Point FFT Using Decimation-in-Time

Given the input sequence $x(n)$ representing a rectangular waveform as in Exercise 6.1, the output sequence $X(k)$, using the DIT flow graph in Figure 6.10, is the same as in Exercise 6.1. The twiddle constants are the same as in Exercise 6.1. Note that the twiddle constant W is multiplied with the second term only (not with the first).

1. **At stage 1:**

$$x(0) + W^0 x(4) = 1 + 0 = 1 \rightarrow x'(0)$$
$$x(0) - W^0 x(4) = 1 - 0 = 1 \rightarrow x'(4)$$

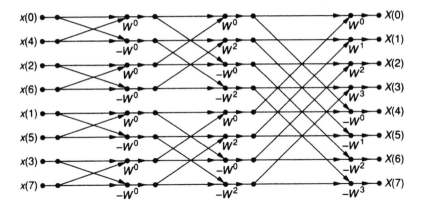

FIGURE 6.10 Eight-point FFT flow graph using decimation-in-time.

$$x(2) + W^0x(6) = 1 + 0 = 1 \rightarrow x'(2)$$
$$x(2) - W^0x(6) = 1 - 0 = 1 \rightarrow x'(6)$$
$$x(1) + W^0x(5) = 1 + 0 = 1 \rightarrow x'(1)$$
$$x(1) - W^0x(5) = 1 - 0 = 1 \rightarrow x'(5)$$
$$x(3) + W^0x(7) = 1 + 0 = 1 \rightarrow x'(3)$$
$$x(3) - W^0x(7) = 1 - 0 = 1 \rightarrow x'(7)$$

where the sequence x's represents the intermediate output after the first iteration and becomes the input to the subsequent stage.

2. At stage 2:

$$x'(0) + W^0x'(2) = 1 + 1 = 2 \rightarrow x''(0)$$
$$x'(4) + W^2x'(6) = 1 + (-j) = 1 - j \rightarrow x''(4)$$
$$x'(0) - W^0x'(2) = 1 - 1 = 0 \rightarrow x''(2)$$
$$x'(4) - W^2x'(6) = 1 - (-j) = 1 + j \rightarrow x''(6)$$
$$x'(1) + W^0x'(3) = 1 + 1 = 2 \rightarrow x''(1)$$
$$x'(5) + W^2x'(7) = 1 + (-j)(1) = 1 - j \rightarrow x''(5)$$
$$x'(1) - W^0x'(3) = 1 - 1 = 0 \rightarrow x''(3)$$
$$x'(5) - W^2x'(7) = 1 - (-j)(1) = 1 + j \rightarrow x''(7)$$

where the intermediate second-stage output sequence x''s becomes the input sequence to the final stage.

3. At stage 3:

$$X(0) = x''(0) + W^0 x''(1) = 4$$

$$X(1) = x''(4) + W^1 x''(5) = 1 - j2.414$$

$$X(2) = x''(2) + W^2 x''(3) = 0$$

$$X(3) = x''(6) + W^3 x''(7) = 1 - j0.414$$

$$X(4) = x''(0) - W^0 x''(1) = 0$$

$$X(5) = x''(4) - W^1 x''(5) = 1 + j0.414$$

$$X(6) = x''(2) - W^2 x''(3) = 0$$

$$X(7) = x''(6) - W^3 x''(7) = 1 + j2.414$$

which is the same output sequence as found in Example 6.1.

6.5 BIT REVERSAL FOR UNSCRAMBLING

A bit-reversal procedure allows a scrambled sequence to be reordered. To illustrate this bit-swapping process, let $N = 8$, represented by three bits. The first and third bits are swapped. For example, $(100)_b$ is replaced by $(001)_b$. As such, $(100)_b$ specifying the address of $X(4)$ is replaced by or swapped with $(001)_b$ specifying the address of $X(1)$. Similarly, $(110)_b$ is replaced/swapped with $(011)_b$, or the addresses of $X(6)$ and $X(3)$ are swapped. In this fashion, the output sequence in Figure 6.5 with the DIF, or the input sequence in Figure 6.10 with the DIT, can be reordered.

This bit-reversal procedure can be applied for larger values of N. For example, for $N = 64$, represented by six bits, the first and sixth bits, the second and fifth bits, and the third and fourth bits are swapped.

Bit Reversal with Indirect Addressing

Swapping memory locations is not necessary if the bit-reversed addressing mode available on the TMS320C3x is used. Let $N = 8$ to illustrate this indirect addressing mode with reversed carry. Given a set of data $x(0), x(1), x(2), \ldots, x(7)$ that we wish to resequence or scramble, to obtain $x(0), x(4), x(2), x(6), x(1), x(5), x(3), x(7)$ as we would do in an FFT using the decimation-in-time (DIT) flow graph in figure 6.10.

1. Set the index register IR0 to one-half the length of the FFT, or IR0 = $N/2$ = 4, assuming a set of real-input sequence. For a complex input sequence, IR0 is set to N to accomodate for the real and imaginary components.

2. Let an auxiliary register such as AR1 contain a base address such as zero or $(0000)_b$ for illustration purpose.

3. The instruction

$$NOP \quad *AR1++(IR0)B$$

is an indirect mode of addressing instruction for bit reversal, introduced in Chapter 2. On execution, the address 0 is selected, then AR1 is incremented to point at memory address 4, which is the base address of zero offset by IR0.

4. On the second execution of this instruction, memory address 4 is selected, then AR1 is incremented to point at the address 2. We arrive at this address by adding the current address to $N/2$, or $(0100)_b + (0100)_b = (0010)_b$ with reversed carry. That is, the carry is to the right, or in the reversed direction, so that the binary addition of 1 and 1 is 0, with a carry of 1 to the right. This is caused by the B in the instruction.

5. On the third execution, memory address 2 is selected, then AR1 is incremented to point to memory address 6, and after the fourth execution, AR1 points to memory address 1, because $(0110)_b + (0100)_b = (0001)_b$ with reversed carry, and so on.

We have used this indirect mode of addressing with reversed carry on the input sequence. We can use a similar procedure on the output sequence, which can be performed by loading the auxiliary register AR1 with the last or highest address, then postdecrementing, or

$$NOP \quad *AR1--(IR0)B$$

This procedure can be used for higher-order FFT length. For a complex FFT, the real components of the input sequence can be arranged in even-numbered addresses and the imaginary components in odd-numbered addresses. The index (offset) register IR0 = N (instead of $N/2$). The programming FFT examples included later incorporate the bit reversal procedure for swapping addresses.

6.6 DEVELOPMENT OF THE FFT ALGORITHM WITH RADIX-4

A radix-4 (base 4) algorithm can increase the execution speed of the FFT. FFT programs on higher radices and split radices have been developed. We will use a decimation-in-frequency (DIF) decomposition process to introduce the development of the radix-4 FFT. The last or lowest decomposition of a radix-4 algorithm consists of four inputs and four outputs. The order or length of the FFT is 4^M, where M is the number of stages. For a 16-point FFT, there are only two stages or iterations as compared with four stages with the radix-2 algorithm.

The DFT in (6.1) is decomposed into four summations, instead of two, as follows:

$$X(k) = \sum_{n=0}^{(N/4)-1} x(n)W^{nk} + \sum_{n=N/4}^{(N/2)-1} x(n)W^{nk} + \sum_{n=N/2}^{(3N/4)-1} x(n)W^{nk} + \sum_{n=3N/4}^{N-1} x(n)W^{nk}$$

$$(6.30)$$

Let $n = n + N/4$, $n = n + N/2$, $n = n + 3N/4$ in the second, third, and fourth summations, respectively. Then (6.30) can be written as

$$X(k) = \sum_{n=0}^{(N/4)-1} x(n)W^{nk} + W^{kN/4}\sum_{n=0}^{(N/4)-1} x(n+N/4)W^{nk}$$

$$+ W^{kN/2}\sum_{n=0}^{(N/4)-1} x(n+N/2)W^{nk} + W^{3kN/4}\sum_{n=0}^{(N/4)-1} x(n+3N/4)W^{nk} \quad (6.31)$$

which represents four $(N/4)$-point DFT's. Using

$$W^{kN/4} = (e^{-j2\pi/N})^{kN/4} = e^{-jk\pi/2} = (-j)^k$$

$$W^{kN/2} = e^{-jk\pi} = (-1)^k$$

$$W^{3kN/4} = (j)^k$$

(6.31) becomes

$$X(k) = \sum_{n=0}^{(N/4)-1} [x(n) + (-j)^k x(n+N/4) + (-1)^k x(n+N/2) + (j)^k x(n+3N/4)]W^{nk}$$

$$(6.32)$$

Let $W_N^4 = W_{N/4}$. Equation (6.32) can be written as,

$$X(4k) = \sum_{n=0}^{(N/4)-1} [x(n) + x(n+N/4) + x(n+N/2) + x(n+3N/4)]W_{N/4}^{nk} \quad (6.33)$$

$$X(4k+1) = \sum_{n=0}^{(N/4)-1} [x(n) - jx(n+N/4) - x(n+N/2) + jx(n+3N/4)]W_N^n W_{N/4}^{nk} \quad (6.34)$$

$$X(4k+2) = \sum_{n=0}^{(N/4)-1} [x(n) - x(n+N/4) + x(n+N/2) - x(n+3N/4)]W_N^{2n} W_{N/4}^{nk} \quad (6.35)$$

$$X(4k+3) = \sum_{n=0}^{(N/4)-1} [x(n) + jx(n+N/4) - x(n+N/2) - jx(n+3N/4)]W_N^{3n} W_{N/4}^{nk} \quad (6.36)$$

for $k = 0, 1, \ldots, (N/4) - 1$. Equations (6.33) through (6.36) represent a decomposition process yielding four four-point DFT's. The flow graph for a 16-point

TABLE 6.1 Twiddle constants for 16-point FFT with radix-4

m	W_N^m	$W_{N/4}^m$
0	1	1
1	$0.9238 - j0.3826$	$-j$
2	$0.707 - j0.707$	-1
3	$0.3826 - j0.9238$	$+j$
4	$0 - j$	1
5	$-0.3826 - j0.9238$	$-j$
6	$-0.707 - j0.707$	-1
7	$-0.9238 - j0.3826$	$+j$

radix-4 decimation-in-frequency FFT is shown in Figure 6.11. Note the four-point butterfly in the flow graph. The $\pm j$ and -1 are not shown in Figure 6.11. The results shown in the flow graph are for the following exercise.

Exercise 6.4 16-Point FFT With Radix-4

Given the input sequence $x(n)$ as in Exercise 6.2, representing a rectangular sequence $x(0) = x(1) = \ldots = x(7) = 1$, and $x(8) = x(9) = \ldots = x(15) = 0$. We will find the output sequence for a 16-point FFT with radix-4 using the flow graph in Figure 6.11. The twiddle constants are shown in Table 6.1.

The intermediate output sequence after stage 1 is shown in Figure 6.11. For example, after stage 1:

$$[x(0) + x(4) + x(8) + x(12)]W^0 = 1 + 1 + 0 + 0 = 2 \rightarrow x'(0)$$

$$[x(1) + x(5) + x(9) + x(13)]W^0 = 1 + 1 + 0 + 0 = 2 \rightarrow x'(1)$$

$$\vdots \qquad\qquad\qquad\qquad \vdots$$

$$[x(0) - jx(4) - x(8) + jx(12)]W^0 = 1 - j - 0 - 0 = 1 - j \rightarrow x'(4)$$

$$\vdots \qquad\qquad\qquad\qquad \vdots$$

$$[x(3) - x(7) + x(11) - x(15)]W^6 = 0 \rightarrow x'(11)$$

$$[x(0) + jx(4) - x(8) - jx(12)]W^0 = 1 + j - 0 - 0 = 1 + j \rightarrow x'(12)$$

$$\vdots \qquad\qquad\qquad\qquad \vdots$$

$$[x(3) + jx(7) - x(11) - jx(15)]W^9 = [1 + j - 0 - 0](-W^1)$$
$$= -1.307 - j0.541 \rightarrow x'(15)$$

For example, after stage 2:

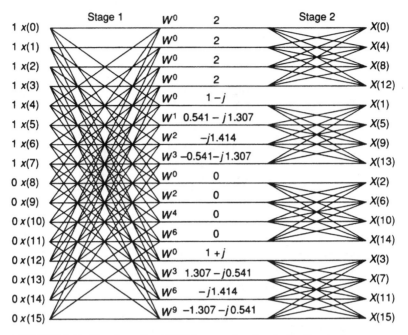

FIGURE 6.11 16-point radix-4 FFT flow graph using decimation-in-frequency.

$$X(3) = (1 + j) + (1.307 - j0.541) + (-j1.414) + (-1.307 - j0.541) = 1 - j1.496$$

and

$$X(15) = (1 + j)(1) + (1.307 - j0.541)(-j) + (-j1.414)(1)$$
$$+ (-1.307 - j0.541)(-j) = 1 + j5.028$$

The output sequence $X(0), X(1), \ldots, X(15)$ is identical to the output sequence obtained with the 16-point FFT with the radix-2 in Figure 6.6. These results also can be verified with MATLAB, described in Appendix B.

The output sequence is scrambled and needs to be resequenced or reordered. This can be done using a digit reversal procedure, in a similar fashion as a bit reversal in a radix-2 algorithm. The radix-4 (base 4) uses the digits 0, 1, 2, 3. For example, the addresses of $X(8)$ and $X(2)$ need to be swapped because $(8)_{10}$ in base 10 or decimal is equal to $(20)_4$ in base 4. Digits 0 and 1 are reversed to yield $(02)_4$ in base 4, which is also $(02)_{10}$ in decimal.

Although mixed or higher radices can provide further reduction in computation, programming considerations become more complex. As a result, the radix-2 is still the most widely used, followed by the radix-4.

6.7 INVERSE FAST FOURIER TRANSFORM

The inverse discrete Fourier transform (IDFT) converts a frequency-domain sequence $X(k)$ into an equivalent sequence $x(n)$ in the time domain. It is defined as

$$x(n) = \frac{1}{N} \sum_{k=0}^{N-1} X(k) W^{-nk} \qquad n = 0, 1, \dots, N-1 \qquad (6.37)$$

Comparing (6.37) with the DFT equation definition in (6.1), we see that the FFT algorithm (forward) described previously can be used to find the IFFT (reverse), with the two following changes:

1. add a scaling factor of $1/N$
2. replace W^{nk} by its complex conjugate W^{-nk}

With the changes, the same FFT flow graphs can be used for the inverse fast Fourier transform (IFFT).

The support tools included with the DSK package contain FFT programming applications. We will also develop programming examples to illustrate the FFT.

A variant of the FFT, such as the fast Hartley transform (FHT) can be obtained readily from the FFT. Conversely, the FFT can be obtained from the FHT [10,11]. A development of the fast Hartley transform (FHT) with flow graphs and exercises for 8 and 16 points FHT's can be found in [12].

Exercise 6.5 Eight-Point IFFT

Let the output sequence $X(0) = 4$, $X(1) = 1 - j2.41, \dots, X(7) = 1 + j2.41$ obtained in Exercise 6.1 become the input to an 8-point IFFT flow graph. Make the two changes (scaling and complex conjugate of W) to obtain an 8-point IFFT (reverse) flow graph from an 8-point FFT (forward) flow graph. The resulting flow graph becomes an IFFT flow graph similar to Figure 6.5. Verify that the resulting output sequence is $x(0) = 1, x(1) = 1, \dots, x(7) = 0$, which represents the rectangular input sequence in Exercise 6.1.

6.8 PROGRAMMING EXAMPLES USING C AND TMS320C3x CODE

We will illustrate the FFT with the following three programming examples using C and TMS320C3x code:

1. A main program that calls an FFT function, both in C code. The resulting output sequence is verified using a simulation procedure

2. A main program in C that calls a real-valued input FFT function in TMS320C3x code, using a simulation procedure

3. A main program in C that calls the same real-valued input FFT function, for a real-time implementation.

Example 6.1 Eight-Point Complex FFT Using C Code

With this programming example, the results are stored in memory, and can be verified. It illustrates a complex FFT with $N = 8$, using a decimation-in-frequency procedure with radix-2. Figure 6.12 shows the main program FFT8C.C in C code that calls a generic FFT function FFT.C, also in C code. The input sequence, specified in the main program, represents a rectangular sequence, $x(0) = \ldots = x(3) = 1000 + j0$ and $x(4) = \ldots = x(7) = 0 + j0$. The main program passes to the FFT function the address of the input data and the FFT length. The header file COMPLEX.H contains the complex structure definition.

The generic FFT function FFT.C is listed in Figure 6.13. The header file TWIDDLE.H included in the FFT function contains the twiddle constants W that allows for an FFT up to 512 points. Different values for W, depending on N, are selected with the variable step in the FFT function. The program TWID-GEN.C (on disk) generates the twiddle constants for a complex FFT. It is to be compiled with Turbo C++ or Borland C++. The resulting file TWIDDLE.H con-

```
/*FFT8C.C - 8-POINT COMPLEX FFT PROGRAM. CALLS FFT.C          */
#include "complex.h"        /*complex structure definition    */
extern void FFT();          /*FFT function                    */
volatile int *out_addr=(volatile int *)0x809802; /*out addr*/

main()
 {
 COMPLEX y[8]={1000,0,1000,0,1000,0,1000,0,
       0.0,0.0,0.0,0.0,0.0,0.0,0.0,0.0}; /*rectangular input*/
 int i, n = 8;
 FFT(y,n);                          /*calls generic FFT function*/
 for (i = 0; i<n; i++)
   {
   *out_addr++ = (y[i]).real;  /*real output component       */
   *out_addr++ = (y[i]).imag;  /*imaginary output component*/
   }
 }
```

FIGURE 6.12 Eight-point FFT program in C that calls a generic FFT function (FFT8C.C).

```
/*FFT.C - FFT RADIX-2 USING DIF. FOR UP TO 512 POINTS     */
#include "complex.h"  /*complex structure definition        */
#include "twiddle.h"  /*header file with twiddle constants*/

void FFT(COMPLEX *Y, int N) /*input sample array, # of points    */
 {
 COMPLEX temp1,temp2;      /*temporary storage variables        */
 int i,j,k;                /*loop counter variables             */
 int upper_leg, lower_leg; /*index of upper/lower butterfly leg */
 int leg_diff;             /*difference between upper/lower leg  */
 int num_stages=0;         /*number of FFT stages, or iterations */
 int index, step;          /*index and step between twiddle factor*/
 i=1;                 /* log(base 2) of # of points = # of stages */
 do
  {
  num_stages+=1;
  i=i*2;
  } while (i!=N);

 leg_diff=N/2; /*starting difference between upper & lower legs*/
 step=512/N;   /*step between values in twiddle.h              */
 for (i=0;i<num_stages;i++)  /*for N-point FFT                 */
  {
  index=0;
  for (j=0;j<leg_diff;j++)
   {
   for (upper_leg=j;upper_leg<N;upper_leg+=(2*leg_diff))
    {
    lower_leg=upper_leg+leg_diff;
    temp1.real=(Y[upper_leg]).real + (Y[lower_leg]).real;
    temp1.imag=(Y[upper_leg]).imag + (Y[lower_leg]).imag;
    temp2.real=(Y[upper_leg]).real - (Y[lower_leg]).real;
    temp2.imag=(Y[upper_leg]).imag - (Y[lower_leg]).imag;
    (Y[lower_leg]).real=temp2.real*(w[index]).real-temp2.imag*(w[index]).imag;
    (Y[lower_leg]).imag=temp2.real*(w[index]).imag+temp2.imag*(w[index]).real;
    (Y[upper_leg]).real=temp1.real;
    (Y[upper_leg]).imag=temp1.imag;
    }
   index+=step;
```

(continued on next page)

FIGURE 6.13 Generic FFT function in C called from a C program (FFT.C).

```
 }
leg_diff=leg_diff/2;
step*=2;
 }
j=0;
for (i=1;i<(N-1);i++)  /*bit reversal for resequencing data*/
 {
  k=N/2;
  while (k<=j)
   {
   j=j-k;
   k=k/2;
   }
  j=j+k;
  if (i<j)
  {
  temp1.real=(Y[j]).real;
  temp1.imag=(Y[j]).imag;
  (Y[j]).real=(Y[i]).real;
  (Y[j]).imag=(Y[i]).imag;
  (Y[i]).real=temp1.real;
  (Y[i]).imag=temp1.imag;
  }
 }
 return;
 }
```

FIGURE 6-13   (continued)

tains 256 sets of complex constant values for W, allowing for an FFT of up to $N = 512$.

From the FFT function, consider the following, with $N = 8$ (see also the 8-point FFT flow graph in Figure 6.5).

1. The loop counter variable $i = 0$ represents the first stage or iteration. The value leg_diff = 4 specifies the difference between the upper and the lower butterfly legs. For example, at stage 1 (first iteration), the operations $y(0) + y(4)$ and $y(0) - y(4)$ are performed, where $y(0)$ and $y(4)$ are designated by upper_leg and lower_leg, respectively. This is an in-place FFT, in which case the memory locations that store the input data samples are again used to store the intermediate and, subsequently, the final output data.

For example, $temp1 = y(0) + y(4) \to y(0)$ and $temp2 = y(0) - y(4) \to y(4)$. The calculation of $y(4)$ after the first stage involves complex operations with the complex twiddle constant W, of the form $(A + jB)(C + jD) = (AC - BD) + j(BC + AD)$, where $j = \sqrt{-1}$, and the constant W can be represented by $C + jD$, with a real and an imaginary component. These calculations are performed with the counter variable $j = 0$. When $j = 1$, `upper_leg` and `lower_leg` specify $y(1)$ and $y(5)$, respectively. Then, $temp1 = y(1) + y(5) \to y(1)$ and $temp2 = y(1) - y(5)$. When $j = 2$, $y(2) + y(6) \to y(2)$. With $j = 3$, $temp1 = y(3) + y(7) \to y(3)$.

The calculations of $y(5)$, $y(6)$, and $y(7)$ after the first stage contain complex operations with the constant W. The variable `index` in `W[index]` represents the W's.

2. The loop counter $i = 1$ represents the second stage, and `leg_diff = 2`. With $j = 0$, `upper_leg` and `lower_leg` specify $y(0)$ and $y(2)$, respectively. The intermediate output results $y(0)$ and $y(2)$ are calculated in a similar manner as in step 1. Then, `upper_leg` and `lower_leg` specify $y(4)$ and $y(6)$, respectively. With $j = 1$ they specify $y(1)$ and $y(3)$, then $y(5)$ and $y(7)$. The intermediate results after stage 2 are then obtained.

3. The loop counter variable $i = 2$ represents the third and final stage, and `leg_diff = 1`. The variable `upper_leg` and `lower_leg` specify $y(0)$ and $y(1)$, respectively. Then, they specify $y(2)$ and $y(3)$, then $y(4)$ and $y(5)$, and finally $y(6)$ and $y(7)$. For each set of values in `upper_leg` and `lower_leg`, similar calculations are performed to obtain the final output from stage 3.

4. The last section in the FFT function performs the bit-reversal procedure and produces a proper sequencing of the output data.

If you have the floating-point tools, compile each program, then link them with the linker command file `FFT8C.CMD` (on the accompanying disk) to create the executable file `FFT8C.OUT` (also on the disk). Download and run `FFT8C.OUT` on the DSK. The output sequence of 16 values, representing real and imaginary components, start at the memory address `809802`. Display these results in decimal with the debugger command

```
memd 0x809802
```

Verify that this is the same output sequence, scaled by 1000 as that obtained in Exercise 6.1 for the 8-point FFT.

Example 6.2 Eight-Point FFT with Real-Valued Input, Using Mixed C and TMS320C3x Code

This example illustrates a real-valued input FFT, as opposed to the more general complex FFT. The input must be real. The resulting output is still complex. In this case, computational requirements can be reduced. The real-valued input FFT can be executed in about half the time as the more general complex FFT.

Figure 6.14 shows a listing of the C program `FFT8MC.C` that calls a real-valued FFT function `FFT_RL.ASM` in TMS320C3x code (on the accompanying disk). This example tests the FFT function using an eight-point FFT. In the next example, we will illustrate the same FFT function with a higher order for a real-time implementation.

The input sequence is real and represents a rectangular waveform with $x(0) = x(1) = x(2) = x(3) = 1000$ and $x(4) = x(5) = x(6) = x(7) = 0$.

```
/*FFT8MC.C - 8-POINT REAL-VALUED FFT. CALLS FFT_RL.ASM IN C3X CODE*/
#include "math.h"
#define N 8                                  /*FFT length  */
#define M 3                                  /*# of stages */
float data[N] = {1,1,1,1,0,0,0,0};    /*real-valued input samples*/
float real1, img1;
extern void fft_rl(int, int, float *);      /*generic FFT function*/
volatile int *IO_OUT = (volatile int *) 0x809802;  /*starting out addr*/

main()
{
  int loop;
  fft_rl(N, M, (float *)data);
  *IO_OUT++ = (int)(data[0]*1000);          /*    XR(0)        */
  for (loop = 1; loop < N/2; loop++)
  {
    real1 = data[loop];
     img1 = data[N-loop];
    *IO_OUT++ = (int)(real1*1000);          /*XR(1)-XR(3)    */
    *IO_OUT++ = (int)(img1*1000);           /*XI(1)-XI(3)    */
  }
  *IO_OUT++ = (int)(data[N/2]*1000);        /*    XR(4)       */
  for (loop = N/2+1; loop < N; loop++)
  {
    real1 = data[N-loop];
     img1 = data[loop];
    *IO_OUT++ = (int)(real1*1000);          /*XR(5)-XR(7)    */
    *IO_OUT++ = (int)(img1*(-1000));        /*XI(5)-XI(7)    */
  }
}
```

FIGURE 6.14 Eight-point FFT program in C that calls a generic real-valued input FFT function (`FFT8MC.C`).

Figure 6.15 shows a listing of the twiddle constants TWID8.ASM for an 8-point real-valued input FFT. Only sine values are shown. When a cosine value is needed, the FFT_RL.ASM function steps through the sine values in TWID8.ASM to obtain the equivalent cosine value. Figure 6.16 shows a C program SINEGEN.C that generated the twiddle constants in Figure 6.15, defining N to be 8 and opening/creating an output file twid8.asm to contain the twiddle constants.

The function FFT_RL.ASM is listed in [9] and is based on the Fortran version in [13]. The bit reversal, performed by the FFT function FFT_RL.ASM, is done on the input sequence. To ensure that the data is properly aligned, a few instructions have been added within the bit-reversal routine in the function FFT_RL.ASM. The changes were made based on a design tip in [14]. Otherwise, the circular buffer used with the bit-reversal procedure would need to be aligned within the main C program [15]. See also reference [16] for an updated version of the real-valued input FFT.

With a real input sequence $x(n)$, the output sequence $X(k) = X_R(k) + jX_I(k)$ is such that:

$$X_R(k) = X_R(N-k) \qquad k = 1, 2, \ldots, N/2 - 1$$
$$X_I(k) = -X_I(N-k) \qquad k = 1, 2, \ldots, N/2 - 1$$
$$X_I(0) = X_I(N/2) = 0 \qquad\qquad (6.38)$$

These conditions are met in Example 6.1, because the imaginary components of the input sequence are zero. Based on the FFT function FFT_RL.ASM, the memory arrangement of the output sequence follows [9]:

$$X_R(0)$$
$$X_R(1)$$
$$\vdots$$
$$X_R(N/2) = X_R(4)$$

```
;TWID8.ASM - TWIDDLE CONSTANTS FOR REAL-VALUED FFT
            .global    _sine
            .data
_sine       .float     0.000000
            .float     0.707107
            .float     1.000000
            .float     0.707107
```

FIGURE 6.15 Twiddle constants for eight-point real-valued input FFT (TWID8.ASM).

```
/*SINEGEN.C - GENERATES SINE VALUES FOR REAL-VALUED INPUT FFT*/
#include <math.h>
#include <stdio.h>
#define N 8
#define pi 3.141592654

main()
{
FILE *stream;
int n;
float result;
stream = fopen("twid8.asm", "w+");
fprintf(stream, "\n%s", "           .global    _sine");
fprintf(stream, "\n%s", "           .data");
fprintf(stream, "\n%s%7f", "_sine    .float     ", 0.0000000);
for (n = 1; n < N/2; n++)
  {
  result = sin(n*2*pi/N);
    fprintf(stream, "\n%s%7f", "              .float     ", result);
  }
fclose(stream);
}
```

FIGURE 6.16. Twiddle constant generator program for real-valued input FFT
(SINEGEN.C).

$$X_I(N/2 - 1) = X_R(3)$$
$$X_I(N/2 - 2) = X_R(2)$$
$$\vdots$$
$$X_I(1) \tag{6.39}$$

Using (6.38), the output sequence in (6.39) is shown in memory in the following order:

$$X_R(0), X_R(1), X_I(1), X_R(2), X_I(2), X_R(3), X_I(3), X_R(4),$$
$$X_R(5), X_I(5), X_R(6), X_I(6), X_R(7), X_I(7)$$

Note that $X_I(0) = X_I(4) = 0$ from (6.38).

Download and run the executable file FFT8MC.OUT into the DSK. The output sequence consists of 14 values (with $X_I(0) = X_I(4) = 0$) in memory locations

`809802-80980f`. Display these values in decimal with the `memd` debugger command, and verify the same results (scaled by 1000) as in Exercise 6.1.

Example 6.3 Real-Time 128-point FFT Using Mixed Code

This example is a real-time implementation version of the previous example, making use of the on-board AIC. Figure 6.17 shows a listing of the main program `FFT128C.C` that calls the same real-valued input FFT function `FFT_RL.ASM` as in the previous example. It performs a 128-point FFT on a real input.

The interrupt rate is set for 10 kHz with the AIC data configuration. A scheme with two buffers allows for a pointer to be switched from one buffer to another, and is more efficient than switching the data from one buffer to another buffer.

To generate the twiddle constants for a 128-point real-valued input FFT, make the following two changes in the program `SINEGEN.C` listed in Figure 6.16: change N to 128, and choose an appropriate filename such as `twid128.asm` to contain the resulting twiddle constants. A total of $N/2$ points (one-half of a sine sequence) or twiddle factors need to be generated for an N-point real-valued input FFT. These twiddle constants are in a format to optimize the execution speed (at a slight cost of memory size). Compile `SINEGEN.C` (with Turbo C++, for example) and execute it to obtain the 64 twiddle constants stored in the file `twid128.asm`. This file needs to be assembled before linking to create the executable COFF file `FFT128C.OUT` (on the accompanying disk). The linker command file is similar to the one in the previous example.

Download and run `FFT128C.OUT` on the DSK. Input to the DSK a 3-kHz sinusoidal signal. The resulting output is displayed in Figure 6.18, obtained from an HP signal analyzer, plotted in the time domain. A similar plot can be obtained from an oscilloscope. Figure 6.18 shows a spike or delta function at the frequency of the input sinusoid. A second delta function represents the folded frequency.

The distance between the two negative spikes corresponds to the sampling frequency of 10 kHz. The negative spike is produced by the main program with the last statement:

```
IO_buffer[0] = -2048
```

This negative spike is used as a reference or range, and is repeated after every frame. The sampling period $T = 1/F_s = 0.1$ ms is the distance between each output sample point. Since there are 128 points ($N = 128$), the distance between the start of each frame, which is the distance between the negative spikes, is 128 × 0.1 ms = 12.8 ms, as can be verified in Figure 6.18. Note that the middle-point between the two spikes corresponds to 5 kHz.

```
/*FFT128C.C - REAL-VALUED FFT WITH 128 POINTS. CALLS FFT_RL.ASM      */
#include "math.h"                             /*standard library func  */
#include "aiccomc.c"                          /*AIC comm routines      */
#define N 128                                 /*size of FFT            */
#define M 7                                   /*number of stages       */
volatile int index = 0;                       /*input_output index     */
float *IO_buffer, *data, *temp;               /*-> array buffers       */
int AICSEC[4] = {0x162C,0x1,0x3872,0x67};     /*AIC config data        */
extern void fft_rl(int, int, float *);        /*fft function protype   */


void c_int05()                                /*interrupt handler func */
{
 PBASE[0x48] = ((int)(IO_buffer[index])) << 2;          /*output data*/
 IO_buffer[index] = (float)(PBASE[0x4C] << 16 >> 18);   /*input data */
 if (++index >= N) index = 0;                 /*increment index, reset = N*/
}


main()
{
 int loop;                                    /* declare variable       */
 float real, img;                             /* declare variables      */
 AICSET_I();                                  /*config AIC for interrupt */
 IO_buffer = (float *) calloc(N, sizeof(float)); /*input_out buffer   */
 data = (float *) calloc(N, sizeof(float));   /* fft data buffer        */
 while (1)                                    /* create endless loop    */
 {
  fft_rl(N, M, (float *)data);                /*call FFT function       */
  data[0] = sqrt(data[0]*data[0])/N;          /*magnitude of X(0)       */
  for (loop = 1; loop < N/2; loop++)          /*calculate X(1)..X(N/2-1) */
  {
   real = data[loop];                         /*real part               */
   img = data[N-loop];                        /*imaginary part          */
   data[loop] = sqrt(real*real+img*img)/N;    /*find magnitude          */
  }
  data[N/2] = sqrt(data[N/2]*data[N/2])/N;    /*magnitude of X(N/2)     */
  for (loop = N/2+1; loop < N; loop++)        /*X(N/2+1).. X(N-1)       */
   data[loop] = data[N-loop];                 /*use symmetry            */
  while (index);                              /*wait till IO_buffer empty */
  temp = data;                                /*temp => data buffer     */
  data = IO_buffer;                           /*IO_buffer->data buffer  */
  IO_buffer = temp;                           /*data buffer->new IO_buffer*/
  IO_buffer[0] = -2048;                       /*sync pulse,negative spike */
 }
}
```

FIGURE 6.17 128-point FFT main program that calls real-valued input FFT function (FFT128C.C).

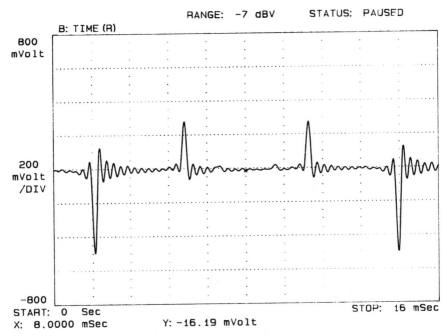

RANGE: −7 dBV   STATUS: PAUSED

FIGURE 6.18 Plot of 128-point real-valued input FFT.

6.9 EXPERIMENT 6: FFT IMPLEMENTATION

The following require the TMS320 floating-point assembly language tools that include a C compiler, an assembler, and a linker. Furthermore, the C program SINEGEN.C that generates twiddle constants needs to be compiled using Turbo C++ or Borland C++.

1. Implement a 16-point complex FFT. Modify the main C program FFT8C.C in Example 6.1 using 16 sets of input samples with COMPLEX y[16] and $n = 16$, in the program. No changes are required in the generic FFT function FFT.C or in the header file TWIDDLE.H that contains the twiddle constants. With a rectangular input sequence as in Example 6.2, verify that the resulting output sequence is the same as in Example 6.2.

2. Implement a 16-point real-valued input FFT with a main program similar to the program FFT8MC.C in Example 6.2 that calls the generic FFT assembly function FFT_RL.ASM. No changes are necessary in the FFT function. Use Example 6.2 or MATLAB, described in Appendix B, to verify your results.

3. Implement a 64-point real-time version of Example 6.3. Use the same scheme as in Example 6.3 to test your results.

REFERENCES

1. J. W. Cooley and J. W. Tukey, "An Algorithm for the Machine Calculation of Complex Fourier Series," *Math. of Computation, 19,* 297–301 (1965).

2. J. W. Cooley, "How the FFT Gained Acceptance," *IEEE Signal Processing Magazine,* pp.10–13, Jan. 1992.

3. J. W. Cooley, "The Structure of FFT and Convolution Algorithms, from a Tutorial," in *IEEE 1990 International Conference on Acoustics, Speech, and Signal Processing,* April 1990.

4. C. S. Burrus and T. W. Parks, *DFT/FFT and Convolution Algorithms: Theory and Implementation,* Wiley, New York, 1988.

5. G. D. Bergland, "A guided tour of the fast Fourier transform," *IEEE Spectrum, 6,* 41–51 (1969).

6. E. O. Brigham, *The Fast Fourier Transform,* Prentice-Hall, Englewood Cliffs, NJ, 1974.

7. S. Winograd, "On Computing the Discrete Fourier Transform," *Math. of Computation,* **32,** 175–199 (1978).

8. H. F. Silverman, "An Introduction to Programming the Winograd Fourier Transform Algorithm (WFTA)," *IEEE Trans. on Acoustics, Speech, and Signal Processing,* **ASSP-25,** 152–165, April (1977).

9. P. E. Papamichalis ed., *Digital Signal Processing Applications with the TMS320 Family—Theory, Algorithms, and Implementations,* Vol.3, Texas Instruments, Inc., Dallas, TX, 1990.

10. R. N. Bracewell, "Assessing the Hartley Transform," *IEEE Trans. on Acoustics, Speech, and Signal Processing,* **ASSP-38,** 2174–2176 (1990).

11. R. N. Bracewell, *The Hartley Transform,* Oxford University Press, New York, 1986.

12. R. Chassaing, *Digital Signal Processing with C and the TMS320C30,* Wiley, New York, 1992.

13. H. V. Sorensen, D. L. Jones, M. T. Heidman, and C. S. Burrus, "Real-Valued Fast Fourier Transform Algorithms," *IEEE Trans. on Acoustics, Speech, and Signal Processing,* **ASSP-35,** 849–863 (1987).

14. *Details on Signal Processing,* Texas Instruments, Inc., Dallas, TX, Fall 1990.

15. *Details on Signal Processing,* Texas Instruments, Inc., Dallas, TX, Winter 1992.

16. *TMS320C3x General-Purpose Applications User's Guide,* Texas Instruments, Inc., Dallas, TX, 1998.

17. P. M. Embree and B. Kimble, *C Language Algorithms for Digital Signal Processing,* Prentice-Hall, Englewood Cliffs, NJ, 1990.

18. S. Kay and R. Sudhaker, "A Zero Crossing Spectrum Analyzer," *IEEE Trans. on Acoustics, Speech, and Signal Processing,* **ASSP-34,** 96–104 February (1986).

19. P. Kraniauskas, "A plain Man's Guide to the FFT," *IEEE Signal Processing Magazine,* April 1994.

20. J. R. Deller, Jr., "Tom, Dick, and Mary Discover the DFT," *IEEE Signal Processing Magazine,* April 1994.

7

Adaptive Filters

- Adaptive structures
- The least mean square (LMS) algorithm
- Programming examples using C and TMS320C3x code

Adaptive filters are best used in cases where signal conditions or system parameters are slowly changing and the filter is to be adjusted to compensate for this change. The least mean square (LMS) criterion is a search algorithm that can be used to provide the strategy for adjusting the filter coefficients. Programming examples are included to give a basic intuitive understanding of adaptive filters.

7.1 INTRODUCTION

In conventional FIR and IIR digital filters, it is assumed that the process parameters to determine the filter characteristics are known. They may vary with time, but the nature of the variation is assumed to be known. In many practical problems, there may be a large uncertainty in some parameters because of inadequate prior test data about the process. Some parameters might be expected to change with time, but the exact nature of the change is not predictable. In such cases, it is highly desirable to design the filter to be self-learning, so that it can adapt itself to the situation at hand.

The coefficients of an adaptive filter are adjusted to compensate for changes in input signal, output signal, or system parameters. Instead of being rigid, an adaptive system can learn the signal characteristics and track slow changes. An adaptive filter can be very useful when there is uncertainty about the characteristics of a signal or when these characteristics change.

Figure 7.1 shows a basic adaptive filter structure in which the adaptive filter's output y is compared with a desired signal d to yield an error signal e, which is fed back to the adaptive filter. The coefficients of the adaptive filter are

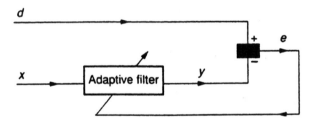

FIGURE 7.1 Basic adaptive filter structure.

adjusted, or optimized, using a least mean square (LMS) algorithm based on the error signal.

We will discuss here only the LMS searching algorithm with a linear combiner (FIR filter), although there are several strategies for performing adaptive filtering.

The output of the adaptive filter in Figure 7.1 is

$$y(n) = \sum_{k=0}^{N-1} w_k(n)x(n-k) \qquad (7.1)$$

where $w_k(n)$ represent N weights or coefficients for a specific time n. The convolution equation (7.1) was implemented in Chapter 4 in conjunction with FIR filtering. It is common practice to use the terminology of weights w for the coefficients associated with topics in adaptive filtering and neural networks.

A performance measure is needed to determine how good the filter is. This measure is based on the error signal,

$$e(n) = d(n) - y(n) \qquad (7.2)$$

which is the difference between the desired signal $d(n)$ and the adaptive filter's output $y(n)$. The weights or coefficients $w_k(n)$ are adjusted such that a mean squared error function is minimized. This mean squared error function is $E[e^2(n)]$, where E represents the expected value. Since there are k weights or coefficients, a gradient of the mean squared error function is required. An estimate can be found instead using the gradient of $e^2(n)$, yielding

$$w_k(n+1) = w_k(n) + 2\beta e(n)x(n-k) \qquad k = 0, 1, \ldots, N-1 \qquad (7.3)$$

which represents the LMS algorithm [1–3]. Equation (7.3) provides a simple but powerful and efficient means of updating the weights, or coefficients, without the need for averaging or differentiating, and will be used for implementing adaptive filters.

The input to the adaptive filter is $x(n)$, and the rate of convergence and accuracy of the adaptation process (adaptive step size) is β.

For each specific time n, each coefficient, or weight, $w_k(n)$ is updated or replaced by a new coefficient, based on (7.3), unless the error signal $e(n)$ is zero. After the filter's output $y(n)$, the error signal $e(n)$ and each of the coefficients $w_k(n)$ are updated for a specific time n, a new sample is acquired (from an ADC) and the adaptation process is repeated for a different time. Note that from (7.3), the weights are not updated when $e(n)$ becomes zero.

The linear adaptive combiner is one of the most useful adaptive filter structures and is an adjustable FIR filter. Whereas the coefficients of the frequency-selective FIR filter discussed in Chapter 4 are fixed, the coefficients, or weights, of the adaptive FIR filter can be adjusted based on a changing environment such as an input signal. Adaptive IIR filters (not discussed here) also can be used. A major problem with an adaptive IIR filter is that its poles may be updated during the adaptation process to values outside the unit circle, making the filter unstable.

The programming examples developed later will make use of equations (7.1)–(7.3). In (7.3), we will simply use the variable β in lieu of 2β.

7.2 ADAPTIVE STRUCTURES

A number of adaptive structures have been used for different applications in adaptive filtering.

1. *For noise cancellation.* Figure 7.2 shows the adaptive structure in Figure 7.1 modified for a noise cancellation application. The desired signal d is corrupted by uncorrelated additive noise n. The input to the adaptive filter is a noise n' that is correlated with the noise n. The noise n' could come from the same source as n but modified by the environment. The adaptive filter's output y is adapted to the noise n. When this happens, the error signal approaches the desired signal d. The overall output is this error signal and not the adaptive filter's output y. This structure will be further illustrated with programming examples using both C and TMS320C3x code.

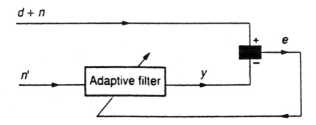

FIGURE 7.2 Adaptive filter structure for noise cancellation.

2. *For system identification.* Figure 7.3 shows an adaptive filter structure that can be used for system identification or modeling. The same input is to an unknown system in parallel with an adaptive filter. The error signal e is the difference between the response of the unknown system d and the response of the adaptive filter y. This error signal is fed back to the adaptive filter and is used to update the adaptive filter's coefficients, until the overall output $y = d$. When this happens, the adaptation process is finished, and e approaches zero. In this scheme, the adaptive filter models the unkown system.

3. Additional structures have been implemented such as:

a) *Notch with two weights,* which can be used to notch or cancel/reduce a sinusoidal noise signal. This structure has only two weights or coefficients, and is illustrated later with a programming example.

b) *Adaptive predictor,* which can provide an estimate of an input. This structure is illustrated later with three programming examples.

c) Adaptive channel equalization, used in a modem to reduce channel distortion resulting from the high speed of data transmission over telephone channels.

The LMS is well suited for a number of applications, including adaptive echo and noise cancellation, equalization, and prediction.

Other variants of the LMS algorithm have been employed, such as the sign-error LMS, the sign-data LMS, and the sign-sign LMS.

1. For the sign-error LMS algorithm, (7.3) becomes

$$w_k(n + 1) = w_k(n) + \beta sgn[e(n)]x(n - k) \qquad (7.4)$$

where *sgn* is the signum function,

$$sgn(u) = \begin{cases} 1 & \text{if } u \geq 0 \\ -1 & \text{if } u < 0 \end{cases} \qquad (7.5)$$

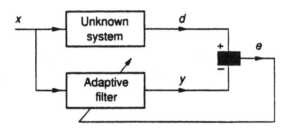

FIGURE 7.3 Adaptive filter structure for system identification.

2. For the sign-data LMS algorithm, (7.3) becomes

$$w_k(n+1) = w_k(n) + \beta e(n)sgn[x(n-k)] \tag{7.6}$$

3. For the sign-sign LMS algorithm, (7.3) becomes

$$w_k(n+1) = w_k(n) + \beta sgn[e(n)]sgn[x(n-k)] \tag{7.7}$$

which reduces to

$$w_k(n+1) = \begin{cases} w_k(n) + \beta & \text{if } sgn[e(n)] = sgn[x(n-k)] \\ w_k(n) - \beta & \text{otherwise} \end{cases} \tag{7.8}$$

which is more concise from a mathematical viewpoint, because no multiplication operation is required for this algorithm.

The implementation of these variants does not exploit the pipeline features of the TMS320C3x processor. The execution speed on the TMS320C3x for these variants can be expected to be slower than for the basic LMS algorithm, due to additional decision-type instructions required for testing conditions involving the sign of the error signal or the data sample.

The LMS algorithm has been quite useful in adaptive equalizers, telephone cancellers, and so forth. Other methods such as the recursive least squares (RLS) algorithm [4], can offer faster convergence than the basic LMS but at the expense of more computations. The RLS is based on starting with the optimal solution and then using each input sample to update the impulse response in order to maintain that optimality. The right step size and direction are defined over each time sample.

Adaptive algorithms for restoring signal properties can also be found in [4]. Such algorithms become useful when an appropriate reference signal is not available. The filter is adapted in such a way as to restore some property of the signal lost before reaching the adaptive filter. Instead of the desired waveform as a template, as in the LMS or RLS algorithms, this property is used for the adaptation of the filter. When the desired signal is available, the conventional approach such as the LMS can be used; otherwise *a priori* knowledge about the signal is used.

7.3 PROGRAMMING EXAMPLES USING C AND TMS320C3x CODE

The following programming examples illustrate adaptive filtering using the least mean square (LMS) algorithm. It is instructive to read the first example even if you have only a limited knowledge of C, since it illustrates the steps in the adaptive process.

Example 7.1 Adaptive Filter Using C Code Compiled With Borland C/C++

This example applies the LMS algorithm using a C-coded program compiled with Borland C/C++. It illustrates the following steps for the adaptation process using the adaptive structure in Figure 7.1:

1. Obtain a new sample for each, the desired signal d and the reference input to the adaptive filter x, which represents a noise signal.
2. Calculate the adaptive FIR filter's output y, applying (7.1) as in Chapter 4 with an FIR filter. In the structure of Figure 7.1, the overall output is the same as the adaptive filter's output y.
3. Calculate the error signal applying (7.2).
4. Update/replace each coefficient or weight applying (7.3).
5. Update the input data samples for the next time n, with a data move scheme used in Chapter 4 with the program FIRDMOVE.C. Such scheme moves the data instead of a pointer.
6. Repeat the entire adaptive process for the next output sample point.

Figure 7.4 shows a listing of the program ADAPTC.C, which implements the LMS algorithm for the adaptive filter structure in Figure 7.1. A desired signal is chosen as $2\cos(2n\pi f/F_s)$, and a reference noise input to the adaptive filter is chosen as $\sin(2n\pi f/F_s)$, where f is 1 kHz, and $F_s = 8$ kHz. The adaptation rate, filter order, number of samples are 0.01, 22, and 40, respectively.

The overall output is the adaptive filter's output y, which adapts or converges to the desired cosine signal d.

The source file was compiled with Borland's C/C++ compiler. Execute this program. Figure 7.5 shows a plot of the adaptive filter's output (y_out) converging to the desired cosine signal. Change the adaptation or convergence rate β to 0.02 and verify a faster rate of adaptation.

Interactive Adaptation

A version of the program ADAPTC.C in Figure 7.4, with graphics and interactive capabilities to plot the adaptation process for different values of β is on the accompanying disk as ADAPTIVE.C, to be compiled with Turbo or Borland C/C++. It uses a desired cosine signal with an amplitude of 1 and a filter order of 31. Execute this program, enter a β value of 0.01, and verify the results in Figure 7.6. Note that the output converges to the desired cosine signal. Press F2 to execute this program again with a different beta value.

```c
//ADAPTC.C - ADAPTATION USING LMS WITHOUT THE TI COMPILER
#include <stdio.h>
#include <math.h>
#define beta 0.01                        //convergence rate
#define N   21                           //order of filter
#define NS  40                           //number of samples
#define Fs  8000                         //sampling frequency
#define pi  3.1415926
#define DESIRED 2*cos(2*pi*T*1000/Fs)    //desired signal
#define NOISE sin(2*pi*T*1000/Fs)        //noise signal

main()
{
 long I, T;
 double D, Y, E;
 double W[N+1] = {0.0};
 double X[N+1] = {0.0};
 FILE *desired, *Y_out, *error;
 desired = fopen ("DESIRED", "w++"); //file for desired samples
 Y_out = fopen ("Y_OUT", "w++");     //file for output samples
 error = fopen ("ERROR", "w++");     //file for error samples
 for (T = 0; T < NS; T++)            //start adaptive algorithm
   {
    X[0] = NOISE;                    //new noise sample
    D = DESIRED;                     //desired signal
    Y = 0;                           //filter'output set to zero
    for (I = 0; I <= N; I++)
     Y += (W[I] * X[I]);             //calculate filter output
    E = D - Y;                       //calculate error signal
    for (I = N; I >= 0; I—)
      {
       W[I] = W[I] + (beta*E*X[I]);  //update filter coefficients
       if (I != 0)
       X[I] = X[I-1];                //update data sample
      }
   fprintf (desired, "\n%10g    %10f", (float) T/Fs, D);
   fprintf (Y_out, "\n%10g    %10f", (float) T/Fs, Y);
   fprintf (error, "\n%10g    %10f", (float) T/Fs, E);
   }
 fclose (desired);
 fclose (Y_out);
 fclose (error);
}
```

FIGURE 7.4 Adaptive filter program compiled with Borland C/C++ (ADAPTC.C).

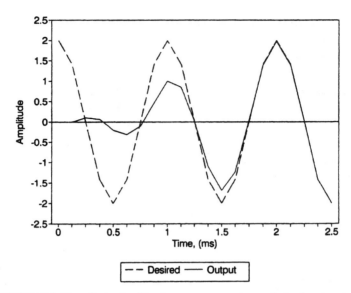

FIGURE 7.5 Plot of adaptive filter's output converging to desired cosine signal.

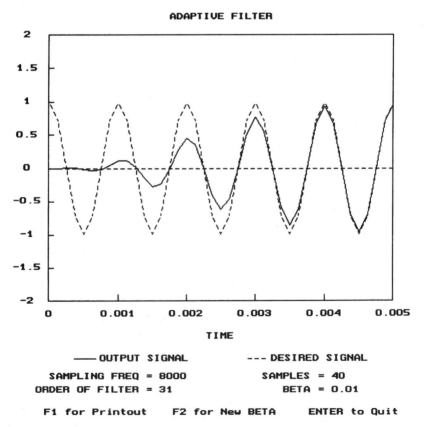

FIGURE 7.6 Plot of adaptive filter's output converging to desired cosine signal using interactive capability with program ADAPTIVE.C.

Example 7.2 Adaptive Filter for Noise Cancellation Using C Code

This example illustrates the adaptive filter structure shown in Figure 7.2 for the cancellation of an additive noise. Figure 7.7 shows a listing of the program ADAPTDMV.C based on the previous program in Example 7.1. Consider the following from the program:

1. The desired signal specified by DESIRED is a sine function with a frequency of 1 kHz. The desired signal is corrupted/added with a noise signal specified by ADDNOISE. This additive noise is a sine with a frequency of 312 Hz. The addition of these two signals is achieved in the program with DPLUSN for each sample period.

2. The reference input to the adaptive FIR filter is a cosine function with a frequency of 312 Hz specified by REFNOISE. The adaptation step or rate of convergence is set to 1.5×10^{-8}, the number of coefficients to 30, and the number of output samples to 128.

3. The output of the adaptive FIR filter y is calculated using the convolution equation (7.1), and converges to the additive noise signal with a frequency of 312 Hz. When this happens, the "error" signal e, calculated from (7.2), approaches the desired signal d with a frequency of 1 kHz. This error signal is the overall output of the adaptive filter structure, and is the difference between the adaptive filter's output y and the primary input consisting of the desired signal with additive noise.

In the previous example, the overall output was the adaptive filter's output. In that case, the filter's output converged to the desired signal. For the structure in this example, the overall output is the error signal and not the adaptive filter's output.

This program was compiled with the TMS320 assembly language floating-point tools, and the executable COFF file is on the accompanying disk. Download and run it on the DSK.

The output can be saved into the file fname with the debugger command

```
save fname,0x809d00,128,L
```

which saves the 128 output samples stored in memory starting at the address 809d00 into the file fname, in ASCII Long format. Note that the desired signal with additive noise samples in DPLUSN are stored in memory starting at the address 809d80, and can be saved also into a different file with the debugger save command.

Figure 7.8 shows a plot of the output converging to the 1-kHz desired sine signal, with a convergence rate of $\beta = 1.5 \times 10^{-8}$. The upper plot in Figure 7.9 shows the FFT of the 1-kHz desired sine signal and the 312-Hz additive noise signal. The lower plot in Figure 7.9 shows the overall output which illustrates the reduction of the 312-Hz noise signal.

```
/*ADAPTDMV.C - ADAPTIVE FILTER FOR NOISE CANCELLATION    */
#include "math.h"
#define beta 1.5E-8              /*rate of convergence     */
#define N 30                     /*# of coefficients       */
#define NS 128                   /*# of output sample points*/
#define Fs  8000                 /*sampling frequency      */
#define pi   3.1415926
#define DESIRED 1000*sin(2*pi*T*1000/Fs) /*desired signal */
#define ADDNOISE 1000*sin(2*pi*T*312/Fs) /*additive noise */
#define REFNOISE 1000*cos(2*pi*T*312/Fs) /*reference noise*/
main()
{
 int I,T;
 float Y, E, DPLUSN;
 float W[N+1];
 float Delay[N+1];
 volatile int *IO_OUTPUT= (volatile int*) 0x809d00;
 volatile int *IO_INPUT = (volatile int*) 0x809d80;
 for (T = 0; T < N; T++)
  {
    W[T] = 0.0;
    Delay[T] = 0.0;
  }
 for (T=0; T < NS; T++)              /*# of output samples    */
  {
    Delay[0] = REFNOISE;        /*adaptive filter's input*/
    DPLUSN = DESIRED + ADDNOISE; /*desired + noise, d+n   */
    Y = 0;
    for (I = 0; I < N; I++)
      Y += (W[I] * Delay[I]);   /*adaptive filter output */
    E = DPLUSN - Y;               /*error signal           */
    for (I = N; I > 0; I—)
    {
      W[I] = W[I] + (beta*E*Delay[I]); /*update weights   */
      if (I != 0)
      Delay[I] = Delay[I-1];            /*update samples   */
    }
    *IO_OUTPUT++ = E;                    /*overall output E */
    *IO_INPUT++ = DPLUSN;                /* store d + n     */
  }
}
```

FIGURE 7.7 Adaptive filter program for sinusoidal noise cancellation using data move (ADAPTDMV.C).

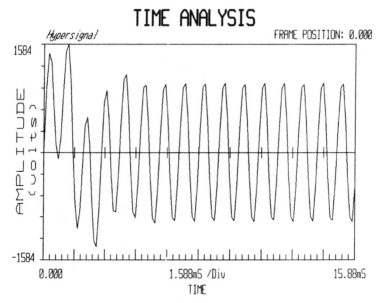

FIGURE 7.8 Plot of overall output of adaptive filter structure converging to 1-kHz desired signal.

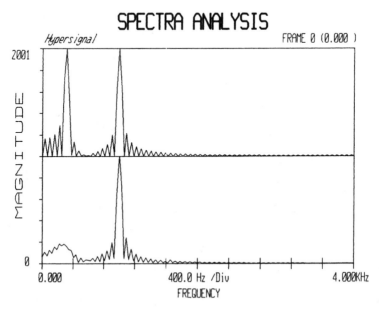

FIGURE 7.9 Output frequency response of adaptive filter structure showing reduction of 312-Hz additive sinusoidal noise.

Examine the effects of different values for the adaptation rate β and for the number of weights or coefficients.

Example 7.3 Adaptive Predictor Using C Code

This example implements the adaptive predictor structure shown in Figure 7.10, with the program ADAPTSH.C shown in Figure 7.11. The input to the adaptive structure is a 1-kHz sine defined in the program. The input to the adaptive filter with 30 coefficients is the delayed input, and the adaptive filter's output is the overall output of the predictor structure.

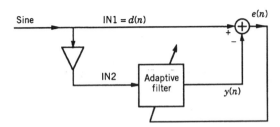

FIGURE 7.10 Adaptive predictor structure.

```
/*ADAPTSH.C - ADAPTIVE FILTER WITH SHIFTED INPUT    */
#include "math.h"
#define beta 0.005          /*rate of convergence */
#define N   30              /*# of coefficients    */
#define NS 128              /*# of output samples */
#define pi   3.1415926
#define shift 90            /*desired amount of shift*/
#define Fs 8000             /*sampling frequency    */
#define inp 1000*sin(2*pi*T*1000/Fs) /*input signal*/

main()
{
```

(continued on next page)

FIGURE 7.11 Adaptive predictor program with arccosine and arcsine for delay (ADAPTSH.C).

```
int I, T;
double xin, x, ys, D, E, Y1;
double W[N+1];
double Delay[N+1];
volatile int *IO_OUTPUT = (volatile int*) 0x809d00;
ys = 0;
for (T = 0; T < N; T++)
 {
  W[T] = 0.0;
  Delay[T] = 0.0;
 }
for (T=0; T < NS; T++) /*# of output samples        */
 {
  xin = inp/1000;      /*input between 1 and -1      */
  if (ys >= xin)       /*is signal rising or falling */
   x = acos(xin);      /*signal is falling          */
  else                 /*otherwise                  */
   x=asin(xin)-(pi/2); /*signal is rising           */
  x = x - (shift);     /*shift                      */
  Delay[0]=cos(x);     /*shifted output=filter's input*/
  D = inp;             /*input data                 */
  Y1 = 0;              /*init output                */
  ys = xin;            /*store input value          */
  for (I=0; I <N; I++) /*for N coefficients         */
   Y1+=W[I]*Delay[I];  /*adaptive filter output     */
  E = D - Y1;          /*error signal               */
  for (I=N; I>0; I—)
   {
    W[I]=W[I]+(beta*E*Delay[I]); /*update weights    */
    if (I != 0)
     Delay[I] = Delay[I-1];      /*update delays     */
   }
  *IO_OUTPUT++ = Y1;             /*overall output    */
 }
}
```

FIGURE 7.11 *(continued)*

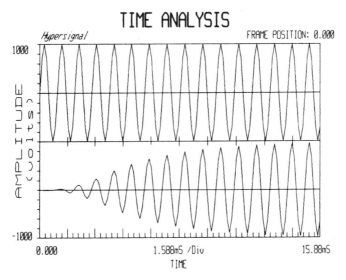

FIGURE 7.12 Output of adaptive predictor converging to desired 1-kHz input signal.

A shifting technique is employed within the program to obtain a delay of 90°. An optimal choice of the delay parameter is discussed in [5]. Note that another separate input is not needed. This shifting technique uses an arccosine or arcsine function depending on whether the signal is rising or falling.

The program SHIFT.C (on disk) illustrates a 90° phase shift. A different amount of delay can be verified with the program SHIFT.C. The program ADAPTSH.C incorporates the shifting section of code.

Verify Figure 7.12, which shows the output of the adaptive predictor (lower graph) converging to the desired 1-kHz input signal (upper graph). When this happens, the error signal converges to zero. Note that 128 output sample points can be collected starting at memory address 809d00.

The following example illustrates this phase shift technique using a table lookup procedure, and Example 7.5 implements the adaptive predictor with TMS320C3x code.

Example 7.4 Adaptive Predictor With Table Lookup for Delay, Using C Code

This example implements the same adaptive predictor of Figure 7.10 with the program ADAPTTB.C listed in Figure 7.13. This program uses a table lookup procedure with the arccosine and arcsine values set in the file scdat (on the accompanying disk) included in the program. The arccosine and arcsine values are selected depending on whether the signal is falling or rising. A delay of 270° is set in the program. This alternate implementation is faster (but not as clean).

```
/*ADAPTTB.C - ADAPTIVE FILTER USING ASIN, ACOS TABLE*/
#define beta 0.005          /*rate of adaptation      */
#define N   30              /*order of filter         */
#define NS 128              /*number of samples       */
#define Fs 8000             /*sampling frequency      */
#define pi 3.1415926
#define inp 1000*sin(2*pi*T*1000/Fs)  /*input        */
#include "scdat"            /*table for asin, acos    */
#include "math.h"

main()
{
 int I, J, T, Y;
 double E, yo, xin, out_data;
 double W[N+1];
 double Delay[N+1];
 volatile int *IO_OUTPUT = (volatile int*) 0x809d00;
 yo=0;
 for (T=0; T < N; T++)
  {
   W[T] = 0.0;
   Delay[T] = 0.0;
  }
 for (T=0; T < NS; T++) /*# of output samples               */
  {
   xin = inp/1000;        /*scale for range between 1 and -1*/
   Y = ((xin)+1)*100;     /*step up array between 0 and 200 */
   if (yo > xin)          /*is signal falling or rising     */
    Delay[0] = yc[Y];     /*signal is falling, acos domain  */
   else                   /*otherwise                       */
    Delay[0] = ys[Y];     /*signal is rising, asin domain   */
   out_data = 0;          /*init filter output to zero      */
    yo = xin;             /*store input                     */
   for (I=0; I<=N; I++)
    out_data +=(W[I]*Delay[I]);    /*filter output          */
   E = xin - out_data;            /*error signal            */
   for (J=N; J > 0; J—)
    {
     W[J]=W[J]+(beta*E*Delay[J]);  /*update coefficients */
     if (J != 0)
     Delay[J] = Delay[J-1];       /*update data samples */
    }
   *IO_OUTPUT++ = out_data*1000;   /* output signal         */
  }
}
```

FIGURE 7.13 Adaptive predictor program with table lookup for delay (ADAPTTB.C).

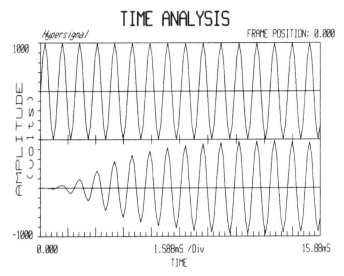

FIGURE 7.14 Output of adaptive predictor (with table lookup procedure) converging to desired 1-kHz input signal.

Figure 7.14 shows the output of the adaptive predictor (lower graph) converging to the desired 1-kHz input signal (upper graph), yielding the same results as in Figure 7.12.

Example 7.5 Adaptive Notch Filter With Two Weights, Using TMS320C3x Code

The adaptive notch structure shown in Figure 7.15 illustrates the cancellation of a sinusoidal interference, using only two weights or coefficients. This structure is discussed in References 1 and 3. The primary input consists of a desired signal d with additive sinusoidal interference noise n. The reference input to the adaptive FIR filter consists of $x_1(n)$, and $x_2(n)$ as $x_1(n)$ delayed by 90°. The output of the two-coefficient adaptive FIR filter is $y(n) = y_1(n) + y_2(n) = w_1(n)x_1(n)$

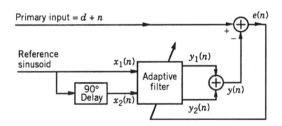

FIGURE 7.15 Adaptive notch structure with two weights.

$+ w_2(n)x_2(n)$. The error signal $e(n)$ is the difference between the primary input signal and the adaptive filter's output, or $e(n) = (d + n) - y(n)$.

Figure 7.16 shows a listing of the program NOTCH2W.ASM, which implements the two-weight adaptive notch filter structure. Consider the following.

```
;NOTCH2W.ASM - ADAPTIVE NOTCH FILTER WITH TWO WEIGHTS
          .start    ".text",0x809900 ;starting address for text
          .start    ".data",0x809C00 ;starting address for data
          .include  "dplusna"    ;data file d+n (1000+312 Hz)
          .include  "cos312a"    ;input data x1(n)
          .include  "sin312a"    ;input data x2(n)
DPN_ADDR  .word     DPLUSN       ;d+n sine 100 + 312 Hz
COS_ADDR  .word     COS312       ;start address of x1(n)
SIN_ADDR  .word     SIN312       ;start address of x2(n)
OUT_ADDR  .word     0x809802     ;output address
SC_ADDR   .word     SC           ;address for sine+cosine samples
WN_ADDR   .word     COEFF        ;address of coefficient w(N-1)
ERF_ADDR  .word     ERR_FUNC     ;address of error function
ERR_FUNC  .float    0            ;init error function to zero
BETA      .float    0.75E-7      ;rate of adaptation
LENGTH    .set      2            ;length of filter N = 2
NSAMPLE   .set      128          ;number of output samples
COEFF     .float    0, 0         ;two weights or coefficients
          .brstart  "SC_BUFF",8  ;align samples buffer
SC        .sect     "SC_BUFF"    ;circular buffer for sine/cosine
          .loop     LENGTH       ;actual length of 2
          .float    0            ;init to zero
          .endloop               ;end of loop
          .entry    BEGIN        ;start of code
          .text                  ;assemble into text section
BEGIN     LDP       WN_ADDR      ;init to data page 128
          LDI       @DPN_ADDR,AR3 ;sin1000 + sin312 addr   -> AR3
          LDI       @COS_ADDR,AR2 ;address of cos312 data
          LDI       @SIN_ADDR,AR5 ;address of sin312 data
          LDI       @OUT_ADDR,AR4 ;output address           -> AR4
          LDI       @ERF_ADDR,AR6 ;error function address  -> AR6
          LDI       LENGTH,BK    ;FIR filter length        -> BK
```

(continued on next page)

FIGURE 7.16 Program listing for adaptive notch filter with two weights (NOTCH2W.ASM).

```
          LDI    @WN_ADDR,AR0   ;w(N-1) address        -> AR0
          LDI    @SC_ADDR,AR1   ;sine+cosine sample addr->AR1
          LDI    NSAMPLE,R5     ;R5=loop counter for # of samples
LOOP      LDF    *AR2++,R3      ;input cosine312 sample-> R3
          STF    R3,*AR1++%     ;store cos sample in SC buffer
          LDF    *AR5++,R3      ;input sine312 sample-> R3
          STF    R3,*AR1++%     ;store sine sample in SC buffer
          LDF    *AR3++,R4      ;input d+n=sin1000 + sin312->R4
          LDI    @WN_ADDR,AR0   ;w(N-1) address -> AR0
          CALL   FILT           ;call FIR subroutine FILT
          SUBF3  R0,R4,R0       ;error = DPN - Y -> R0
          FIX    R0,R1          ;convert R0 to integer -> R1
          STI    R1,*AR4++      ;store error as output
          MPYF   @BETA,R0       ;R0=error function=beta*error
          STF    R0,*AR6        ;store error function
          LDI    @WN_ADDR,AR0   ;w(N-1) address -> AR0
          CALL   ADAPT          ;call adaptation routine
          SUBI   1,R5           ;decrement loop counter
          BNZ    LOOP           ;repeat for next sample
WAIT      BR     WAIT           ;wait
;FIR FILTER SUBROUTINE
FILT      MPYF3  *AR0++,*AR1++%,R0 ;w1(n)*x1(n) = y1(n) -> R0
          LDF    0,R2           ;R2 = 0
          MPYF3  *AR0++,*AR1++%,R0 ;w2(n)*x2(n) = y2(n) -> R0
||        ADDF3  R0,R2,R2       ;R2 = y1(n)
          ADDF3  R0,R2,R0       ;y1(n)+y2(n) = y(n)  -> R0
          RETSU                 ;return from subroutine
;ADAPTATION SUBROUTINE
ADAPT     MPYF3  *AR6,*AR1++%,R0 ;error function*x1(n)-> R0
          LDF    *AR0,R3        ;w1(n) -> R3
          MPYF3  *AR6,*AR1++%,R0 ;error function*x2(n)-> R0
||        ADDF3  R3,R0,R2       ;w1(n)+error function*x1(n)->R2
          LDF    *+AR0,R3       ;w2(n) -> R3
||        STF    R2,*AR0++      ;w1(n+1)=w1(n)+error function*x1(n)
          ADDF3  R3,R0,R2       ;w2(n)+error function*x2(n)->R2
          STF    R2,*AR0        ;w2(n+1)=w2(n)+error function*x2(n)
          RETSU                 ;return from subroutine
```

FIGURE 7.16 *(continued)*

1. The desired signal d is chosen to be a sine function with a frequency of 1 kHz and the interference or noise n is a sine function with a frequency of 312 Hz. The data points for these signals were generated with the program ADAPT-DMV.C discussed in Example 7.2. The addition of these two signals represents $d + n$, and the resulting data points are contained in the file dplusna, which is included in the program NOTCH2W.ASM.

2. The first input to the adaptive filter $x_1(n)$ is chosen as a 312-Hz cosine function, and the second input $x_2(n)$ as a 312-Hz sine function. The data points for these two functions are contained in the files cos312a and sin312a, respectively, which are included in the program NOTCH2W.ASM.

3. The error function, defined as $\beta e(n)$, as well as the two weights, are initialized to zero.

4. A circular buffer SC_BUFF of length two is created for the cosine and sine samples. A total of 128 output samples are obtained starting at the address 809802. An input cosine sample is first acquired as $x_1(n)$, with the instruction LDF *AR2++,R3, and stored in the circular buffer; then an input sine sample is acquired as $x_2(n)$, with the instruction LDF *AR5++,R3, and stored in the subsequent memory location in the circular buffer. Then the primary input sample, which represents $d + n$ is acquired, with the instruction LDF *AR3++,R4. This sample in DPLUSN is used for the calculation of the error signal.

5. The filter and adaptation routines are separated in the program in order to make it easier to follow the program flow. For faster execution, these routines can be included where they are called. This would eliminate the CALL and RETS instructions.

6. The LMS algorithm is implemented with equations (7.1)–(7.3) by calculating first the adaptive filter's output $y(n)$, followed by the error signal $e(n)$, and then the two coefficients $w_1(n)$ and $w_2(n)$.

The filter subroutine finds $y(n) = w_1(n)x_1(n) + w_2(n)x_2(n)$, where the coefficients or weights $w_1(n)$ and $w_2(n)$ represent the weights at time n. Chapter 4 contains many examples for implementing FIR filters using TMS320C3x code. In this case, there are only two coefficients and two input samples. The first input sample to the adaptive FIR filter is the cosine sample $x_1(n)$ from the file cos312a, and the second input sample is the sine sample $x_2(n)$ from the file sin312a.

7. The adaptive filter's output sample for each time n is contained in R0. The error signal, for the same specific time n, is calculated with the instruction SUBF3 R0,R4,R0, where R4 contains the sample $d + n$. The first sample of the error signal is not meaningful, because the adaptive filter's output is zero for the first time n. The two weights, initialized to zero, are not yet updated.

8. The error function, which is the product of $e(n)$ in R0, and β, is stored in memory specified by AR6.

9. Within the adaptation routine, the first multiply instruction yields R0, which contains the value $\beta e(n)x_1(n)$ for a specific time n. The second multiply instruction yields R0, which contains the value $\beta e(n)x_2(n)$. This multiply instruction is in parallel with an ADDF3 instruction in order to update the first weight $w_1(n)$.

10. The parallel addition instruction

$$|\,|\qquad \text{ADDF3}\quad \text{R3,R0,R2}$$

adds R3, which contains the first weight $w_1(n)$, and R0, which contains $\beta e(n)x_1(n)$ from the first multiply instruction.

11. The instruction

$$\text{LDF}\qquad \text{*+AR0,R3}$$

loads the second weight $w_2(n)$ into R3. Note that AR0 is preincremented *without* modification to the memory address of the second weight.

12. The instruction

$$|\,|\qquad \text{STF}\quad \text{R2,*AR0++}$$

stores the updated weight R2 $= w_1(n + 1)$ in the memory address specified by AR0. Then, AR0 is postincremented to point at the address of the second weight. The second ADDF3 instruction is similar to the first one and updates the second weight $w_2(n + 1)$.

For each time n, the preceding steps are repeated. New cosine and sine samples are acquired, as well as $(d + n)$ samples. For example, the adaptive filter's output $y(n)$ is calculated with the newly acquired cosine and sine samples and the previously updated weights. The error signal is then calculated and stored as output.

The 128 output samples can be retrieved from memory and saved into a file n2w with the debugger command

```
save n2w,0x809802,128,1
```

The adaptive filter's output $y(n)$ converges to the additive 312-Hz interference noise n. The "error" signal $e(n)$, which is the overall output of this adaptive filter structure, becomes the desired 1-kHz sine signal d. Figure 7.17 shows a plot of the output error signal converging to the desired 1-kHz sine signal d. Reduce slightly the adaptation rate β to 0.5×10^{-7} and verify that the output adapts slower to the 1-kHz desired signal. However, if β is much too small, such as $\beta = 1.0 \times 10^{-10}$, the adaptation process will not be seen, with only 128 output samples and the same number of coefficients.

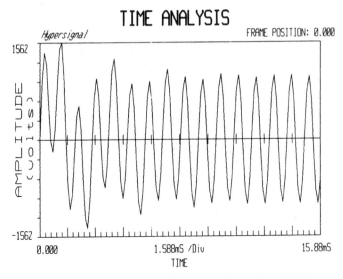

FIGURE 7.17 Output of adaptive notch filter converging to desired 1-kHz sine signal.

Example 7.6 Adaptive Predictor Using TMS320C3x Code

This example implements the adaptive predictor structure shown in Figure 7.10, using TMS320C3x code. The primary input is a desired sine signal. This signal is delayed and becomes the input to the adaptive FIR filter as a cosine signal with the same frequency as the desired signal and one-half its amplitude. The output of the adaptive filter $y(n)$ is adapted to the desired signal $d(n)$.

Figure 7.18 shows a listing of the program ADAPTP.ASM for the adaptive predictor. The desired signal is a 312-Hz sine signal contained in the file sin312a. The input to the 50-coefficient adaptive FIR filter is a 312-Hz cosine signal contained in the file hcos312a. It represents the desired signal delayed with one-half the amplitude. The program ADAPTC.C in Example 7.1 was used to create the two files with the sine and cosine data points. The FIR filter program BP45SIMP.ASM in Appendix B can be instructive for this example. Consider the following.

1. The 50 coefficients or weights of the FIR filter are initialized to zero. The circular buffer XN_BUFF, aligned on a 64-word boundary, is created for the cosine samples. The cosine samples are placed in the circular memory buffer in a similar fashion as was done in Chapter 4 in conjunction with FIR filters. For example, note that the first cosine sample is stored into the last or bottom memory location in the circular buffer.

2. A total of 128 output samples are obtained starting at memory address

```
;ADAPTP.ASM - ADAPTIVE PREDICTOR
            .start    ".text",0x809900    ;starting address for text
            .start    ".data",0x809C00    ;starting address for data
            .include  "sin312a"           ;data for sine of 312 Hz
            .include  "hcos312a"          ;data for 1/2 cosine 312 Hz
            .data                         ;data section
D_ADDR      .word     SIN312              ;desired signal address
HC_ADDR     .word     HCOS312             ;addr of input to adapt filter
OUT_ADDR    .word     0x809802            ;output address
XB_ADDR     .word     XN+LENGTH-1         ;bottom addr of circular buffer
WN_ADDR     .word     COEFF               ;coefficient address
ERF_ADDR    .word     ERR_FUNC            ;address of error function
ERR_FUNC    .float    0                   ;init ERR FUNC to zero
BETA        .float    1.0E-8              ;rate of adaptation
LENGTH      .set      50                  ;FIR filter length
NSAMPLE     .set      128                 ;number of output samples
COEFF       .float    0,0,0,0,0,0,0,0,0,0,0,0,0,0,0,0,0,0,0,0,0,0,0,0,0
            .float    0,0,0,0,0,0,0,0,0,0,0,0,0,0,0,0,0,0,0,0,0,0,0,0,0
            .brstart  "XN_BUFF",64        ;align samples buffer
XN          .sect     "XN_BUFF"           ;circ buffer for filter samples
            .loop     LENGTH              ;buffer size for samples
            .float    0                   ;init samples to zero
            .endloop                      ;end of loop
            .entry    BEGIN               ;start of code
            .text                         ;text section
BEGIN       LDP       XB_ADDR             ;init to data page 128
            LDI       @D_ADDR,AR2         ;desired signal addr   -> AR2
            LDI       @HC_ADDR,AR3        ;1/2 cosine address    -> AR3
            LDI       @OUT_ADDR,AR5       ;output address        -> AR4
            LDI       @ERF_ADDR,AR6       ;error function addr   -> AR6
            LDI       LENGTH,BK           ;FIR filter length     -> BK
            LDI       @WN_ADDR,AR0        ;coeff w(N-1) address -> AR0
            LDI       @XB_ADDR,AR1        ;bottom of circ buffer-> AR1
            LDI       NSAMPLE,R5          ;R5=loop counter for # samples
LOOP        LDF       *AR3++,R3           ;input to adapt filter-> R3
            STF       R3,*AR1++(1)%       ;store @ bottom of circ buffer
            LDF       *AR2++,R4           ;input desired sample -> R4
            LDI       @WN_ADDR,AR0        ;w(N-1) address-> AR0
            CALL      FILT                ;call FIR routine
```

(continued on next page)

FIGURE 7.18 Program listing for adaptive predictor (ADAPTP.ASM).

```
        FIX       R0,R1              ;convert Y to integer -> R1
        STI       R1,*AR5++          ;store to output memory buffer
        SUBF3     R0,R4,R0           ;error = D-Y -> R0
        MPYF      @BETA,R0           ;R0=ERR FUNC=beta*error
        STF       R0,*AR6            ;store error function
        LDI       LENGTH-2,RC        ;reset repeat counter
        LDI       @WN_ADDR,AR0       ;w(N-1) address -> AR0
        CALL      ADAPT              ;call adaptation routine
        SUBI      1,R5               ;decrement loop counter
        BNZ       LOOP               ;repeat for next sample
WAIT    BR        WAIT               ;wait
;FIR FILTER SUBROUTINE
FILT    LDF       0,R2               ;R2 = 0
        RPTS      LENGTH-1           ;repeat LENGTH-1 times
        MPYF3     *AR0++,*AR1++%,R0  ;w(N-1-i)*x(n-(N-1-i))
||      ADDF3     R0,R2,R2           ;accumulate
        ADDF3     R0,R2,R0           ;add last product y(n)->R0
        RETSU                        ;return from subroutine
;ADAPTATION SUBROUTINE
ADAPT   MPYF3     *AR6,*AR1++%,R0    ;ERR FUNC*x(n-(N-1)) -> R0
        LDF       *AR0,R3            ;w(N-1) -> R3
        RPTB      LOOP_END           ;repeat LENGTH-2 times
        MPYF3     *AR6,*AR1++%,R0    ;ERR FUNC*x(n-(N-1-i))->R0
||      ADDF3     R3,R0,R2           ;w(N-1-i)+ERR FUNC*x(n-(N-1-i))
LOOP_END LDF      *+AR0(1),R3        ;load subsequent H(k) -> R3
||      STF       R2,*AR0++          ;store/update coefficient
        ADDF3     R3,R0,R2           ;w(n+1)=w(n)+ERR FUNC*x(n)
        STF       R2,*AR0            ;store/update last coefficient
        RETSU                        ;return from subroutine
```

FIGURE 7.18 *(continued)*

809802. The FIR filter routine FILT and the adaptation routine ADAPT are in separate sections to make it easier to follow the program flow. For faster execution, these routines can be placed where they are called, eliminating the call and return from subroutine intructions.

3. Before the adaptation routine is called to update the weights or coefficients, the repeat counter register RC is initialized with LENGTH $-$ 2, or RC = 48. As a result, the repeat block of code is executed 49 times (repeated 48 times), including the STF R2, *AR0++ instruction in parallel.

Figure 7.19 shows the overall output $y(n)$ converging to the desired 312-Hz

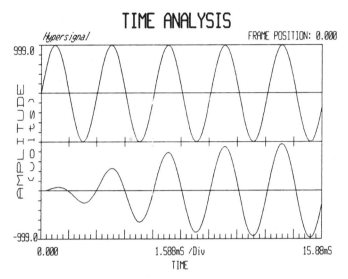

FIGURE 7.19 Output of adaptive predictor (lower plot) converging to desired 312-Hz sine signal.

sine input signal. Reduce the adaptation rate to 10^{-10} and verify a slower rate of adaptation to the 312-Hz sine signal.

Example 7.7 Real-Time Adaptive Filter for Noise Cancellation, Using TMS320C3x Code

This example illustrates the basic adaptive filter structure in Figure 7.2 as an adaptive notch filter. The two previous examples are very useful for this implementation. Two inputs are required in this application, available on the AIC on board the DSK. While the primary input (PRI IN) is through an RCA jack, a second input to the AIC is available on the DSK board from pin 3 of the 32-pin connector JP3.

The secondary or auxiliary input (AUX IN) should be first tested with the loop program (LOOP.ASM) discussed in Chapter 3. Four values are set in AIC-SEC to configure the AIC in the loop program. Replace 0x63 (or 0x67) with 0x73 to enable the AIC auxiliary input and bypass the input filter on the AIC, as described in the AIC secondary communication protocol in Chapter 3. With an input sinusoidal signal from pin 3 of the connector JP3, the output (from the RCA jack) is the delayed input.

Figure 7.20 shows the program listing ADAPTER.ASM for this example.

The AIC configuration data set in AICSEC specifies a sampling rate of 15,782 Hz or ≅ 16 kHz, as can be verified using similar calculations made in the exercises in Chapter 3 to calculate a desired sampling frequency. However,

```
;ADAPTER.ASM-ADAPTIVE STRUCTURE FOR NOISE CANCELLATION. OUTPUT AT e(n)
                .start    ".text",0x809900   ;where text begins
                .start    ".data",0x809C00   ;where data begins
                .include  "AICCOM31.ASM"     ;AIC communications routines
                .data                        ;assemble into data section
AICSEC          .word     162Ch,1h,244Ah,73h ;For AIC,Fs = 16K/2 = 8 kHz
NOISE_ADDR .word          NOISE+LENGTH-1     ;last address of noise samples
WN_ADDR         .word     COEFF              ;address of coefficients w(N-1)
ERF_ADDR        .word     ERR_FUNC           ;address of error function
ERR_FUNC        .float    0                  ;initialize error function
BETA            .float    2.5E-12            ;rate of adaptation constant
LENGTH          .set      50                 ;set filter length
COEFF:                                       ;buffer for coefficients
                .loop     LENGTH             ;loop length times
                .float    0                  ;init coefficients to zero
                .endloop                     ;end of loop
                .brstart  "XN_BUFF",128      ;align buffer for noise samples
NOISE           .sect     "XN_BUFF"          ;section for input noise samples
                .loop     LENGTH             ;loop length times
                .float    0                  ;initialize noise samples
                .endloop                     ;end of loop
                .entry    BEGIN              ;start of code
                .text                        ;assemble into text section
BEGIN           LDP       WN_ADDR            ;init to data page 128
                CALL      AICSET             ;initialize AIC
                LDI       @ERF_ADDR,AR6      ;error function address     ->AR6
                LDI       LENGTH,BK          ;filter length              ->BK
                LDI       @WN_ADDR,AR0       ;coefficient address w(N-1) ->AR0
                LDI       @NOISE_ADDR,AR1    ;last noise sample address  ->AR1
LOOP            CALL      IOAUX              ;get noise sample from AUX IN
                FLOAT     R6,R3              ;transfer noise sample into R3
                STF       R3,*AR1++%         ;store noise sample-> circ buffer
                LDI       @WN_ADDR,AR0       ;w(N-1) coefficients address->AR0
FILT            LDF       0,R2               ;R2 = 0
                RPTS      LENGTH-1           ;next 2 instr (LENGTH-1) times
                MPYF3     *AR0++,*AR1++%,R0  ;w(N-1-i)*x(n-(N-1-i))
||              ADDF3     R0,R2,R2           ;accumulate
                ADDF3     R0,R2,R0           ;add last product=y(n) -> R0
                CALL      IOPRI              ;signal+noise d+n from pri input
```

(continued on next page)

FIGURE 7.20 Real-time adaptive filter program for noise cancellation (ADAPTER.ASM).

```
                    FLOAT    R6,R4              ;R4= d+n in floating-point
                    SUBF3    R0,R4,R0           ;error e => R0 = (d+n)-y
                    FIX      R0,R7              ;R7=R0 in integer
                    MPYF     @BETA,R0           ;R0=ERR FUNC=beta*e
                    STF      R0,*AR6            ;store error function
                    LDI      LENGTH-2,RC        ;set repeat counter register RC
                    LDI      @WN_ADDR,AR0       ;w(N-1) coefficients address->AR0
                    CALL     ADAPT              ;call ADAPT subroutine
                    BR       LOOP               ;repeat with next sample
;ADAPTATION ROUTINE
ADAPT               MPYF3    *AR6,*AR1++%,R0    ;error function*x(n-(N-1))->R0
                    LDF      *AR0,R3            ;w(N-1) -> R3
                    RPTB     LOOP_END           ;repeat length-2 times
                    MPYF3    *AR6,*AR1++%,R0    ;error function*x(n-(N-1-i))->R0
    ||              ADDF3    R3,R0,R2           ;w(N-1-i)+error func*x(n-(N-1-i))
LOOP_END            LDF      *+AR0(1),R3        ;load subsequent w(k) -> R3
    ||              STF      R2,*AR0++          ;store/update coefficient
                    ADDF3    R3,R0,R2           ;w(n+1)=w(n)+error function*x(n)
                    STF      R2,*AR0            ;store/update coefficient
                    RETS                        ;return from subroutine
```

FIGURE 7.20 *(continued)*

in this implementation, the actual sampling rate is one-half that or $\cong$ 8 kHz since *both* inputs on the AIC are accessed. Note that the AIC input bandpass filter is not inserted.

Initially, the 50 coefficients of the adaptive FIR filter are set to zero. A circular buffer XN_BUFF for the noise samples, aligned on a 128-word boundary, is initialized with zero.

The AIC communication routines in AICCOM31.ASM, included in the ADAPTER.ASM program, are set so that the extended precision registers R6 and R7 are used for input and output, respectively. These routines were tested in Chapters 3–5.

Within the block of code starting with the label LOOP, the auxiliary and the primary inputs on the AIC are accessed through the routines IOAUX and IOPRI, respectively. When the routine IOAUX is called, an output is obtained through R7, then a new noise sample is obtained through R6. The FIR filter calculates the output at time *n*. Then the subroutine IOPRI is called, and an output is obtained again through R7 and a new sample $(d + n)$ is acquired from the AIC primary input. For each time *n*, both routines IOPRI and IOAUX are called to acquire a new sample $(d + n)$ from the AIC primary input and a new sample *n* from the AIC reference or auxiliary input, respectively.

The FIR filter code section, starting with FILT, is incorporated directly into the program for faster execution. See also the adaptive notch filter program with two weights NOTCH2W.ASM and the adaptive predictor program ADAPTP.ASM, discussed in the two previous examples. The adaptation routine, starting with the label ADAPT, is kept separately to make the program easier to follow.

The "error" signal e is the overall output that is the difference between the AIC primary input $(d + n)$ and the adaptive filter's output y. As y adapts to n, the "error" signal converges to d.

Test this program:

1. The AIC primary input consists of a desired 1-kHz sinusoidal signal added to another sinusoidal noise signal with a frequency such as 700 Hz.

2. The reference or auxiliary input also consists of the 700-Hz sinusoidal noise. Use a T-connector to input the noise signal to the auxiliary AIC input, as well as to a summer circuit at the same time. Use a passive summer circuit or an OP AMP to add the 1-kHz and the 700-Hz sinusoidal signals.

3. Run the adaptive filter program. Change the frequency of the 700-Hz noise to 600 or 800 Hz.

4. Observe the output from an oscilloscope. Initially the output shows the added sinusoids (1 kHz and 700 Hz), then converges within a few seconds to the 1-kHz desired signal.

5. Change the rate of adaptation, reassemble and run the program again. A much larger β value (by 1000) will not show the adaptation process. The output would immediately display the 1-kHz desired signal. It is more interesting to test this implementation with a smaller value of β (by a factor of 10). Verify a slower rate of adaptation.

This example is extended to adapt continuously using the program ADAPTERC.ASM (on disk). Input a 1-kHz sinusoidal signal s from the AIC primary input (adaptive FIR filter input). Input a desired noise signal n from the AIC auxiliary input. A summer circuit is not needed since (s + n) is performed within the program. Verify that the adaptation process takes place every 4 seconds. The resulting output converges to the desired random noise signal, and the 1-kHz sinusoidal signal is cancelled out.

7.4 EXPERIMENT 7: ADAPTIVE FILTERING IMPLEMENTATION

1. Implement the C-coded simulation Examples 7.1–7.4

2. Implement the TMS320C3x-coded simulation Examples 7.5 and 7.6.

3. Implement the real-time adaptive filter Example 7.7.

4. Implement a simulation version of Example 7.7. Choose a 1-kHz desired sine input signal and a 312-Hz additive and reference sine noise signal. Collect 128 output samples. Illustrate the convergence of the output "error" signal $e(n)$ to the 1-kHz desired signal. The necessary files required for this experiment have been utilized in the adaptive filtering examples. For example, the file dplusna contains the data points of a 1-kHz sine signal added with a 312-Hz sine signal.

5. Implement Example 7.7 using a bandlimited random noise signal as the additive and reference signal. Note that the pseudorandom noise generator program PRNOISE.ASM in Chapter 3, as input to an FIR lowpass filter program, produces a bandlimited random noise signal. This scheme would require two DSK boards, one to provide the bandlimited random noise and the other to implement the adaptive filter structure.

6. Implement a real-time version of the adaptive predictor Example 7.4 in C. Use the table lookup procedure to obtain the delayed input to the adaptive filter. Care should be exercised to set the input to the adaptive filter to a zero DC level. A shifted signal could also be produced with a phase-shifter circuit.

REFERENCES

1. B. Widrow and S. D. Stearns, *Adaptive Signal Processing,* Prentice-Hall, Englewood Cliffs, NJ, 1985.

2. B. Widrow and M. E. Hoff, Jr., "Adaptive Switching Circuits," in *IRE WESCON,* pp. 96–104, 1960.

3. B. Widrow, J. R. Glover, J. M. McCool, J. Kaunitz, C. S. Williams, R. H. Hearn, J. R. Zeidler, E. Dong, Jr., and R. C. Goodlin, "Adaptive Noise Cancelling: Principles and Applications," in *Proceedings of the IEEE,* **63,** 1692–1716 (1975).

4. J. R. Treichler, C. R. Johnson, Jr., and M. G. Larimore, *Theory and Design of Adaptive Filters,* Wiley, New York, 1987.

5. J. R. Zeidler, "Performance Analysis of LMS Adaptive Prediction Filters," *Proc. IEEE,* **78,** 1781–1806 (1990).

6. S. Haykin, *Adaptive Filter Theory,* Prentice-Hall, Englewood Cliffs, NJ, 1986.

7. S. T. Alexander, *Adaptive Signal Processing Theory and Applications,* Springer-Verlag, New York, 1986.

8. C. F. Cowan and P. F. Grant eds., *Adaptive Filters,* Prentice-Hall, Englewood Cliffs, NJ, 1985.

9. M. L. Honig and D. G. Messerschmitt, *Adaptive Filters: Structures, Algorithms and Applications,* Kluwer Academic, Norwell, MA, 1984.

10. V. Solo and X. Kong, *Adaptive Signal Processing Algorithms: Stability and Performance,* Prentice-Hall, Englewood Cliffs, NJ, 1995.

11. S. Kuo, G. Ranganathan, P. Gupta, and C. Chen, "Design and Implementation of Adaptive Filters," *IEEE 1988 International Conference on Circuits and Systems,* June 1988.

12. S. M. Kuo and D. R. Morgan, *Active Noise Control Systems,* Wiley, 1996.

13. P. Papamichalis ed., *Digital Signal Processing Applications with the TMS320 Family— Theory, Algorithms, and Implementations,* Vol. 3, Texas Instruments Inc., Dallas, TX, 1990.

14. M. G. Bellanger, *Adaptive Digital Filters and Signal Analysis,* Marcel Dekker, New York, 1987.

15. R. Chassaing and B. Bitler, "Adaptive Filtering with C and the TMS320C30 Digital Signal Processor," in *Proceedings of the 1992 ASEE Annual Conference,* June 1992.

16. R. Chassaing, D. W. Horning, and P. Martin, "Adaptive Filtering with the TMS320C25," in *Proceedings of the 1989 ASEE Annual Conference,* June 1989.

8

DSP Applications and Projects

This Chapter can be used as a source of experiments, projects, and applications. A wide range of projects have been implemented based on both the floating-point TMS320C30 digital signal processor [1–6], briefly described at the end of this chapter, and the fixed-point TMS320C25 [7]. They range in topics from communications and controls, to neural networks, and can be used as a source of ideas to implement other projects. The proceedings from the TMS320 Educators Conferences, published by Texas Instruments, Inc., contain a number of TMS320-based articles and can be a good source of project ideas [3–5]. Applications described in References 8 and 9 and the previous chapters on filtering and the fast Fourier transform as well as Appendices B–D also can be useful.

8.1 BANKS OF FIR FILTERS

This project implements eight different filters, with eight sets of FIR filter coefficients incorporated into one program. Each set contains 55 coefficients, designed with a sampling frequency of $F_s = 10$ kHz. They represent the following filters:

1. Lowpass with a cutoff frequency of $F_s/4$
2. Highpass with a cutoff frequency of $F_s/4$
3. Bandpass with a center frequency of $F_s/4$
4. Bandstop with a center frequency of $F_s/4$
5. 2-Passbands with center frequencies at 1 and 3 kHz
6. 3-Passbands with center frequencies at 1, 2, and 3 kHz
7. 4-Passbands with center frequencies at 0.5, 1.5, 2.5, and 3.5 kHz
8. 3-Stopbands with center frequencies at 1, 2, and 3 kHz

These FIR filter coefficients were introduced in Chapter 4. Figure 8.1 shows a

```
;FIR8SETP - PARTIAL PROGRAM WITHOUT EIGHT SETS OF COEFFICIENTS
        .start   ".text",0x809900 ;starting address of text
        .start   ".data",0x809C00 ;starting address of data
        .include "AICCOM31.ASM"    ;AIC comm routine
        .data                      ;data section
LENGTH  .set     55                ;# of filter taps
FN      .set     3                 ;set desired filter number
AICSEC  .word    162Ch,1h,3872h,63h ;Fs= 10 kHz
STORE   .word    COEFF             ;starting addr of coeff
COEFF   .word    COEFF1            ;address of 1st set of LP coeff
        .word    COEFF2            ;address of 2nd set of HP coeff
        .word    COEFF3            ;address of 3rd set of BP coeff
        .word    COEFF4            ;address of 4th set of BS coeff
        .word    COEFF5            ;address of 2-passbands coeff
        .word    COEFF6            ;address of 3-passbands coeff
        .word    COEFF7            ;address of 4-passbands coeff
        .word    COEFF8            ;address of 3-stopbands coeff
        .entry   BEGIN             ;start of code
        .text                      ;text section
BEGIN   LDP      AICSEC            ;init to data page 128
        CALL     AICSET            ;init AIC
        LDI      @XN_ADDR,AR1      ;last sample address ->AR1
        LDI      @STORE,AR2        ;start addr of coeff table->AR2
        LDI      LENGTH,BK         ;BK = size of circular buffer
        LDI      FN,IR0            ;IR0 = selected filter number
        SUBI     1,IR0            ;correct OFFSET+1 to OFFSET
LOOP    CALL     AICIO_P           ;AIC I/O routine IN->R6,OUT->R7
        FLOAT    R6,R3             ;input new sample -> R3
        LDI      *+AR2(IR0),AR0    ;selected coefficient set->AR0
        CALL     FILT              ;go to subroutine FILT
        FIX      R2,R7             ;R7=integer(R2)
        BR       LOOP              ;loop continuously
FILT    STF      R3,*AR1++%        ;newest sample to model delay
        LDF      0,R0              ;init R0=0
        LDF      0,R2              ;init R2=0
        RPTS     LENGTH-1          ;multiply LENGTH times
        MPYF3    *AR0++,*AR1++%,R0 ;H(N)*X(N)
||      ADDF3    R0,R2,R2          ;in parallel with accum -> R2
```

(continued on next page)

FIGURE 8.1 Partial program (without coefficients) for implementing eight filters (FIR8SETP).

```
        ADDF    R0,R2              ;last accum -> R2
        RETS                       ;return from subroutine
        .data                      ;coefficients section
;LOWPASS COEFFICIENTS - COEFF1
COEFF1 .float   3.6353E-003,-1.1901E-003,-4.5219E-003, 2.6882E-003
        .
        .
        .

;3-STOPBANDS COEFFICIENTS - COEFF8
COEFF8 .float   3.6964e-002,-1.0804e-001, 1.7129e-001,-1.1816e-001
        .
        .
        .

XN_ADDR .word   XN+LENGTH-1        ;last (newest) input sample
        .brstart "XN_BUFF",64      ;align samples buffer
XN      .sect   "XN_BUFF"          ;section for input samples
        .loop   LENGTH             ;loop length times
        .float  0                  ;init input samples
        .endloop                   ;end of loop
        .end                       ;end
```

FIGURE 8.1 *(continued)*

listing of the program FIR8SETP, which is a partial program that does not include the eight sets of coefficients. The complete program FIR8SETS.ASM is on the accompanying disk, with the eight sets of coefficients incorporated directly in the program. Consider the program in Figure 8.1.

The starting address of each of the eight sets of coefficients is specified with COEFF1, COEFF2, . . . , COEFF8. These eight addresses are contained in a table. The auxiliary register AR0 specifies the starting address of the selected set of coefficients. That address is loaded into AR0 with the instruction LDI *+AR2(IR0),AR0. This loads the starting address of the coefficient table offset by the index register IR0. A value of FN = 3 sets the auxiliary register IR0 to 2 (subtracted by 1) causing the address specified by COEFF3 to be loaded into AR0. This is the starting address of the third set of coefficients in the table, which represents a bandpass filter.

A value of FN = 1 would load the first address COEFF1, which is the starting address of the first set of coefficients, which represents a lowpass filter.

Verify both the bandpass filter with FN = 3 and the lowpass filter with FN = 1.

Interactive Implementation

The program FIR8SETS.ASM (on the accompanying disk) contains the eight sets of filter coefficients and can be made interactive with a C program. The program FIRALL.ASM (on the accompanying disk) is created by making the following changes in FIR8SETS.ASM:

1. The assembler directive FN   .set   3 is replaced with FN   .word 809800h.
2. The instruction LDI FN, IR0 is replaced with the two instructions LDI @FN,AR4 and LDI *AR4,IR0

The program FIRALL.CPP shown in Figure 8.2 is compiled and linked using Borland C/C++. This is done in a similar fashion to the programs PC-COM.CPP and LOOPCOM.CPP, discussed in Chapter 3 in conjunction with the PC host communicating with the TMS320C31 on the DSK. Execute the program FIRALL.EXE (on disk) and enter the selected filter number as shown in the menu from Figure 8.3. The value entered is passed to the assembly-coded

```
//FIRALL.CPP - PROGRAM WHICH INTERACTS WITH FIRALL.ASM
#include "dsklib.h"
void main()
{
 char *msg;  //pointer to any error message if it occurs
 MSGS err;   //enumerated message for looking up messages
 unsigned long hostdata = 0;
 clrscr();
 Detect_Windows();
 Init_Communication(10000);
 HALT_CPU(); // Put C31 into spin0 mode
 clrscr();
 printf("\n\n");
 printf("\n        Filters with 55 coefficients");
 printf("\n\n\n                      1)..........LOWPASS");
 printf("\n                      2).........HIGHPASS");
 printf("\n                      3).........BANDPASS");
 printf("\n                      4).........BANDSTOP");
 printf("\n                      5).........2-PASSBANDS");
 printf("\n                      6).........3-PASSBANDS");
 printf("\n                      7).........4-PASSBANDS");
```

(continued on next page)

FIGURE 8.2 PC host program that interacts with DSK program with 8 sets of coefficients (FIRALL.CPP).

```
printf("\n                           8)..........3-STOPBANDS");
printf("\n\n\n     Select filter number (1-8) : ");
scanf ("%d", &hostdata);
putmem(0x809800L, 1, &hostdata);
if((err=Load_File("firall.dsk",LOAD))==NO_ERR) //load task
  {
  RUN_CPU();
  }
else
  {
  msg = Error_Strg(err);
  printf("\r\n%s",msg);   //print error message if it occurs
  exit(0);
  }
}
```

FIGURE 8.2   *(continued)*

program FIRALL.ASM through memory location 809800 (reserved memory location for the boot loader), from step 1. Then the selected filter number, now in FN, is loaded into the index register IR0 from step 2.

Verify that the selected filter is implemented.

If you simply change a set of 55 coefficients with a different set of 55 coefficients, you can use these two interactive programs to implement different filters. Reassemble only the FIRALL.ASM program. The C program FIRALL.CPP need not be recompiled, since it downloads and runs FIRALL.DSK. You will need to recompile/relink FIRALL.CPP if you add more sets of coefficients and wish to have the appropriate prompts from the C program.

```
Filters with 55 coefficients


              1)..........LOWPASS
              2)..........HIGHPASS
              3)..........BANDPASS
              4)..........BANDSTOP
              5)..........2-PASSBANDS
              6)..........3-PASSBANDS
              7)..........4-PASSBANDS
              8)..........3-STOPBANDS
```

FIGURE 8.3 User selection menu for one of eight types of FIR filters.

Extend this project by making use of the external hardware interrupt circuit, shown in Figure 8.7 and described in Section 8.4, to control the amplitude of a generated sinusoid. Construct the external hardware circuitry and assemble/run the program FIR8EXT.ASM (on disk). An FIR filter with two passbands is implemented since the filter number FN is initialized to 5 in the program. Press the switch in Figure 8.7. This causes an external interrupt to occur, the filter number FN is incremented to 6, and the corresponding filter with three passbands is implemented.

Verify that each time the switch is pressed, the subsequent filter is implemented. After the eighth filter with the three stopbands is implemented, FN is reset to 1 on pressing the switch.

Using this scheme, one can "step through" a sequence of options or events; in this example, the implementation of a series of FIR filters.

8.2 MULTIRATE FILTER

With multirate processing, a filter can be realized with fewer coefficients than an equivalent single-rate approach. Possible applications include a graphic equalizer, a controlled noise source, and background noise synthesis.

You can test this project now by first reading the implementation section.

Introduction

Multirate processing uses more than one sampling frequency to perform a desired processing operation. The two basic operations are decimation, which is a sampling-rate reduction, and interpolation, which is a sampling-rate increase [10–17]. Multirate decimators can reduce the computational requirements of the filter. A sampling-rate increase by a factor of K can be achieved with interpolation by padding or adding $K - 1$ zeros between pairs of consecutive input samples x_i and x_i+1. A noninteger sampling-rate increase or decrease can be obtained by cascading the interpolation process with the decimation process. For example, if a net sampling-rate increase of 1.5 is desired, we would interpolate by a factor of three, padding two zeros between each input sample, and then decimate with the interpolated input samples shifted by two before each calculation. Decimating or interpolating over several stages generally results in better efficiency.

Design Considerations

A binary random signal is fed into a bank of filters that can be used to shape an output spectrum. The functional block diagram of the multirate filter is shown in Figure 8.4. The frequency range is divided into 10 octave bands, with each band being $\frac{1}{3}$-octave controllable. The control of each octave band is achieved

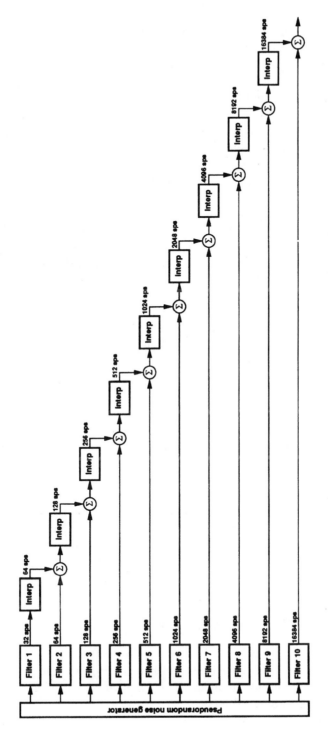

FIGURE 8.4 Functional block diagram of multirate filter with 10 bands.

229

with three filters. The coefficients of these filters are combined to yield a composite filter with one set of coefficients for each octave. Only three unique sets of coefficients (low, middle, and high) are required , because the center frequency and the bandwidth are proportional to the sampling frequency. Each of the $\frac{1}{3}$-octave filters has a bandwidth of approximately 23% of its center frequency, a stopband rejection of greater than 45 dB, with an amplitude that can be controlled individually. This control provides the capability of shaping an output pseudorandom noise spectrum. Forty-one coefficients are used for the highest $\frac{1}{3}$-octave filter to achieve these requirements.

In order to meet the filter specifications in each region with a constant sampling rate, the number of filter coefficients must be doubled from one octave filter to the next-lower one. As a result, the lowest-octave filter would require 41 $\times 2^9$ coefficients. With 10 filters ranging from 41 coefficients to 41×2^9 coefficients, the computational requirements would be considerable. To overcome these computational requirements, the multirate approach shown in Figure 8.4 is implemented.

The noise generator is a software-based implementation of a maximal length sequence technique used for generating pseudorandom numbers, and was introduced in Chapter 3. The output of the noise generator provides uncorrelated noise input to each of the 10 sets of FIR bandpass filters shown in Figure 8.4. In Chapter 3, we developed two program versions of the pseudorandom noise generator, and we also used the generated noise sequence as input to an FIR filter in Chapter 4.

Because each $\frac{1}{3}$-octave filter can be scaled individually, a total of 30 levels can be controlled. The output of each octave bandpass filter, except the last one, becomes the input to an interpolation lowpass filter, using a 2:1 interpolation factor. The ripple in the output spectrum is minimized by having each adjacent $\frac{1}{3}$-octave filter with crossover frequencies at the 3-dB points.

The center frequency and bandwidth of each filter are determined by the sampling rate. The sampling rate of the output is chosen to be 16,384 Hz. The highest-octave filter is processed at 16,384 samples per second, and each successively lower-octave band is processed at half the rate of the next-higher band.

Only three separate sets of 41 coefficients are used for the lower, middle, and higher $\frac{1}{3}$-octave bands. For each octave band, the coefficients are combined as follows:

$$H_{ij} = (H_{lj})(L_{3i-2}) + (H_{mj})(L_{3i-1}) + (H_{hj})(L_{3i})$$

where $i = 1, 2, \ldots, 10$ bands and $j = 0, 1, \ldots, 40$ coefficients. $L_1, L_2, \ldots, L_{30}$ represent the level of each $\frac{1}{3}$-octave band filter, and $H_{lj}, H_{mj},$ and H_{hj} represent the jth coefficient of the lower, middle, and higher $\frac{1}{3}$-octave band FIR filter. For example, for the first band, with $i = 1$

$$H_0 = (H_{l0})(L_1) + (H_{m0})(L_2) + (H_{h0})(L_3)$$

$$H_1 = (H_{l1})(L_1) + (H_{m1})(L_2) + (H_{h1})(L_3)$$

$$\vdots$$

$$H_{40} = (H_{l40})(L_1) + (H_{m40})(L_2) + (H_{h40})(L_3)$$

and for band 10, with $i = 10$

$$H_0 = (H_{l0})(L_{28}) + (H_{m0})(L_{29}) + (H_{h0})(L_{30})$$

$$\vdots$$

$$H_{40} = (H_{l40})(L_{28}) + (H_{m40})(L_{29}) + (H_{h40})(L_{30})$$

For an efficient design, lower-octave bands are processed at a lower sampling rate, then interpolated up by a factor of two to a higher sampling rate, to be summed with the next-higher octave band filter output, as shown in Figure 8.4. Each interpolation filter is a 21-coefficient FIR lowpass filter, with a cutoff frequency of approximately one-fourth of the sampling rate. For each input, the interpolation filter provides two outputs, or

$$y_1 = x_0 I_0 + 0 I_1 + x_1 I_2 + 0 I_3 + \ldots + x_{10} I_{20}$$

$$y_2 = 0 I_0 + x_0 I_1 + 0 I_2 + x_1 I_3 + \ldots + x_9 I_{19}$$

where y_1 and y_2 are the first and second interpolated outputs, respectively, x_n are the filter inputs, and I_n are the interpolation filter coefficients. The interpolator is processed in two sections to provide the data-rate increase by a factor of two.

For the multirate filter, the approximate number of multiplication operations (with accumulation) per second is

$$MAC = (41 + 21)(32 + 64 + 128 + 256 + 512 + 1{,}024 + 2{,}048 + 4{,}096 + 8{,}192)$$

$$+ (41)(16{,}384) \cong 1.68 \times 10^6$$

Note that no interpolation is required for the last stage.

To find the approximate equivalent number of multiplications for the single-rate filter to yield the same impulse response duration, let

$$N_s T_s = N_m T_m$$

where N_s and N_m are, respectively, the number of single-rate and multirate coefficients, and T_s and T_m are, respectively, the single-rate and multirate sampling periods. Then

$$N_s = N_m \frac{F_s}{F_m}$$

where F_s and F_m are, respectively, the single-rate and multirate sampling frequencies. For example, for band 1,

$$N_s = (41)(16{,}384/32) = 20{,}992$$

using F_s as the sampling rate of the highest band, and $F_m = 32$ for the first band. For band 2

$$N_s = (41)(16{,}384/64) = 10{,}496$$

For band 3, $N_s = 5{,}248$; for band 10, $N_s = 41$. The total number of coefficients for the single-rate filter would then be

$$N_s = 20{,}992 + 10{,}496 + \ldots + 41 = 41{,}943$$

The approximate number of multiplications (with accumulation) per second for an equivalent single-rate filter is then

$$MAC = N_s F_s = 687 \times 10^6$$

which would considerably increase the processing time and data storage requirements.

A brief description of the main processing follows, for the first time through.

Band 1

1. Run the bandpass filter and obtain one output sample.
2. Run the lowpass interpolation filter twice and obtain two outputs. The interpolator provides two sample outputs for each input sample.
3. Store in buffer B_2's first two memory locations. Three buffers are utilized in this scheme: buffers B_1 and B_2, each of size 512, and buffer B_3 of size 256.

Band 2

1. Run bandpass filter two times and sum with the two previous outputs stored in buffer B_2, from band 1.
2. Store summed values in the same memory locations of B_2 again.
3. Pass sample in B_2's first memory location to interpolation filter twice and obtain two outputs.
4. Store these two outputs in buffer B_3.
5. Pass sample in B_2's second memory location to interpolation filter twice and obtain two outputs.
6. Store these two outputs in B_3's third and fourth memory locations.

Band 3

1. Run bandpass filter four times and sum with the previous four outputs stored in B_3 from band 2.

2. Store summed values in B_3's first four memory locations.

3. Pass sample in B_3's first memory location to interpolation filter twice and obtain two outputs.

4. Store these two outputs in buffer B_2's first two memory locations.

5. Pass sample in B_3's second memory location to interpolation filter twice and obtain two outputs.

6. Store these two outputs in buffer B_2's third and fourth memory locations.

7. Repeat steps 3 and 4 for the other two samples in B_3's third and fourth memory locations. Store each of these samples, obtain two outputs, and store each set of two outputs in B_2's fifth through eighth memory locations.

⋮

Band 10

1. Run bandpass filter 512 times and sum with the previous 512 outputs stored in B_2, from band 9.

2. Store summed values in B_2's memory locations 1 through 512.

No interpolation is required for band 10. After all the bands are processed, wait for the output buffer B_1 to be empty. Then switch the buffers B_1 and B_2— the last working buffer with the last output buffer. The main processing is then repeated again.

A time of approximately 5.3 ms was measured for the main processing loop using the following scheme.

a) Output to an oscilloscope a positive value set at the beginning of the main processing loop.

b) At the end of the main processing loop, negate the value set in the previous step. Output this negative level to the oscilloscope.

c) The processing time is the duration of the positive level set in step a) and can be measured with the oscilloscope.

The highest sampling rate (in kilosamples per second) is the ratio of the number of samples and the processing time, and is approximately

$$F_s(\max) = \frac{512}{5.3 \text{ ms}} = 96.6 \text{ ksps}$$

Implementation

Test the multirate filter with the program MR7DSK.ASM, which is on the accompanying disk. This program is a 7-band version of the 10-band multirate filter. Only 2K words of internal memory are available on the TMS320C31 with no external memory available on the DSK board. To implement the 10-band version, over 3K words of memory for code and data are required. The 10-band version MR10SRAM.ASM (on the accompanying disk) is implemented using the daughter board with the external memory described in Appendix C.

All the levels or scale values are initialized to zero in MR7DSK.ASM. These levels, L_1–L_{21} are specified in the program by SCALE_1L,   SCALE_1M, SCALE_1U, . . . ,   SCALE_7M,   SCALE_7U, which represent the lower, middle, and upper $\frac{1}{3}$-octave scales for the 7 bands.

Set SCALE_7M (L_{20}) to 1 in order to turn ON the middle $\frac{1}{3}$-octave filter of band 7. The sampling frequency is set for approximately 8 kHz in AICSEC, with the values of A and B as 126Ch and 4892h, respectively (calculated in Chapter 3).

Figure 8.5 shows the frequency response of the middle $\frac{1}{3}$-octave filter of band 7, with a center frequency of one-quarter the sampling frequency, or 2 kHz.

Turn on the middle $\frac{1}{3}$-octave filter of band 6 by setting SCALE_6M to 1, and reinitialize band 7 to zero. Verify a bandpass filter with a center frequency of 1 kHz, which is one-quarter of the effective sampling rate of 4 kHz for band 6. The middle $\frac{1}{3}$-octave of band 5 has a center frequency of 512 Hz, which is one-quarter the effective sampling rate of 2 kHz. Turn on all three $\frac{1}{3}$-octave filters of band 4 (all other bands set to zero) and verify a wider bandwidth bandpass filter with a center frequency of approximately 256 Hz. Note that the middle $\frac{1}{3}$-octave band 1 filter, with SCALE_1M set to 1 and all others to 0, yields a bandpass filter centered at 32 Hz.

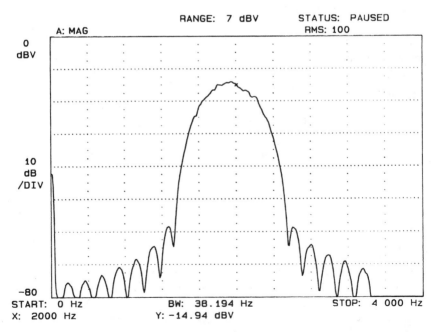

FIGURE 8.5 Frequency response of middle $\frac{1}{3}$-octave of band 7 filter using 7 bands.

Divide the AIC master clock by 16 in the AIC communication program AICCOM31.ASM by changing the instruction LDI 1,R0 to LDI 16,R0. This was illustrated in Chapter 1. Reassemble the program MR7DSK.ASM, not AICCOM31.ASM. The overall sampling rate of the seventh band is approximately 512 Hz, which is one-sixteenth of the originally set sampling rate. Verify that band 1 now yields a bandpass filter centered at 2 Hz. The frequency response of this band 1 filter centered at 2 Hz is as selective or sharp as the filter's response shown in Figure 8.5.

Note that it is possible to obtain bandpass filters centered at frequencies between 1 Hz and one-quarter the sampling rate set for the highest band, by turning ON the appropriate band.

8.3 PASS/FAIL ALARM GENERATOR

An alarm generator can be achieved by generating different tones. Chapter 5 illustrates the generation of a sinusoidal waveform or tone based on the recursive difference equation. Figure 8.6 shows the program ALARMGEN.ASM, which implements this alarm generator.

```
;ALARMGEN.ASM - PASS/FAIL ALARM GENERATOR
        .start  "intsect",0x809fC5  ;starting addr for interrupt
        .start  ".text",0x809900    ;starting addr for text
        .start  ".data",0x809C00    ;starting addr for data
        .include "AICCOM31.ASM"      ;AIC comm routines
        .sect   "intsect"            ;section for interrupt vector
        BR      ISR                  ;XINT0 interrupt vector
        .data                        ;assemble into data section
AICSEC  .word   162Ch,1H,3872h,67H  ;AIC data
TIME_P  .set    3000                ;length of pass signal
TIME_R  .set    2                   ;# of repetitions of fail signal
TIME_F  .set    3000                ;length of fail signal
SEED    .word   7E521603H           ;initial seed value
A1      .float  +1.618034           ;A coefficient for 1-kHz
A2      .float  +0.618034           ;A coefficient for 2-kHz
A4      .float  -1.618034           ;A coefficient for 4-kHz
Y1      .float  +0.5877853          ;C coefficient for 1-kHz
Y2      .float  +0.9510565          ;C coefficient for 2-kHz
```

(continued on next page)

FIGURE 8.6 Program for pass/fail alarm generator (ALARMGEN.ASM).

```
Y4        .float  +0.5877853   ;C coefficient for 4-kHz
B         .float  -1.0         ;B coefficient
Y0        .float  0.0          ;initial condition
SCALER    .float  1000         ;output scaling factor
          .entry  BEGIN        ;start of code
          .text                ;assemble into text section
BEGIN     LDP     AICSEC       ;init to data page 128
          CALL    AICSET_I     ;initialize AIC
          LDI     @SEED,R0     ;R0 = initial seed value
          PUSH    R0           ;save R0 into stack
LOOP_N    POP     R0           ;restore R0 from stack
          LDI     R0,R4        ;load seed into R4
          LSH     -31,R4       ;move bit 31 to LSB    =>R4
          LDI     R0,R2        ;R2 = R0 = SEED
          LSH     -30,R2       ;move bit 30 to LSB    =>R2
          ADDI    R2,R4        ;add bits (31+30)      =>R4
          LDI     R0,R2        ;R2 = R0 = SEED
          LSH     -28,R2       ;move bit 28 to LSB    =>R2
          ADDI    R2,R4        ;add bits (31+30+28)   =>R4
          LDI     R0,R2        ;R2 = R0 = SEED
          LSH     -17,R2       ;move bit 17 to LSB    =>R2
          ADDI    R2,R4        ;add bits(31+30+28+17)=>R4
          AND     1,R4         ;mask LSB of R4
          LSH     1,R0         ;shift SEED left by 1
          OR      R4,R0        ;put R4 intoLSB of R0
          PUSH    R0           ;save R0 into STACK
          LDI     R4,R4        ;store integer R4
          BNZ     LOOP_P       ;to PASS loop if # 0
          BZ      LOOP_F       ;to FAIL loop if 0
;SEQUENCE FOR PASS SIGNAL => 4 kHz (if a 1)
LOOP_P    PUSH    R4           ;store R4 into stack
          LDI     TIME_P,R6    ;length of PASS signal
          PUSH    R6           ;save R6 into stack
          LDF     @Y0,R1       ;initially R1 = Y(0) = 0
          LDF     @Y4,R1       ;initially R1 = Y(1)
          LDF     @A4,R3       ;R3=A
          MPYF3   R3,R1,R1     ;R1=AxY(1)
          LDF     @Y4,R0       ;R0=Y2 (previously Y1) due to delay
          LDF     @B,R4        ;R4=B
```

(continued on next page)

FIGURE 8.6 *(continued)*

```
            BR      WAIT            ;go wait for interrupt
;SEQUENCE FOR 2-kHz FAIL SIGNAL (if a 0)
LOOP_F  PUSH    R4              ;save R4 into stack
        LDI     TIME_R,R6       ;# of repetitions of FAIL signal
        PUSH    R6              ;save R6 into stack
LOOP_F2 LDI     TIME_F,R6       ;length of FAIL signal
        LDF     @Y0,R1          ;initially R1 = Y(0) = 0
        LDF     @Y2,R1          ;initially R1 = Y(1)
        LDF     @A2,R3          ;R3=A
        MPYF3   R3,R1,R1        ;R1=A x Y(1)
        LDF     @Y2,R0          ;R0=Y2 (previously Y1) due to delay
        LDF     @B,R4           ;R4=B
        BR      WAIT            ;go wait for interrupt
;SEQUENCE FOR 1-kHz FAIL SIGNAL (if a 0, after 2-kHz signal)
LOOP_F1 LDI     TIME_F,R6       ;length FAIL signal
        LDF     @Y0,R1          ;initially R1 = Y(0) = 0
        LDF     @Y1,R1          ;initially R1 = Y(1)
        LDF     @A1,R3          ;R3=A
        MPYF3   R3,R1,R1        ;R1=A x Y(1)
        LDF     @Y1,R0          ;R0=Y2 (previously Y1) due to delay
        LDF     @B,R4           ;R4=B
;Y(n) FOR n >= 3
WAIT    IDLE                    ;wait for interrupt
        BR      WAIT            ;branch to wait
;INTERRUPT VECTOR
ISR     LDF     R1,R2           ;R2 = A x Y1
        MPYF3   R3,R1,R1        ;R1 = A(A x Y1)
        MPYF3   R4,R0,R0        ;R0 = B x Y2
        ADDF    R0,R1           ;R1 = output
;OUTPUT ROUTINE
        PUSH    R6              ;save R6
        LDF     R1,R7           ;store R1 into R7
        MPYF    @SCALER,R7      ;scale output
        FIX     R7,R7           ;convert R7 into integer
        CALL    AICIO_I         ;AIC I/O routine,output R7
        LDF     R2,R0           ;R0 = A x Y1  for next n
        POP     R6              ;restore R6
        SUBI    1,R6            ;decrement time counter
        BZ      CONT            ;continue TIME_( ) = 0
```

(continued on next page)

FIGURE 8.6 *(continued)*

```
          RETI                    ;return from interrupt
CONT      POP     R6              ;restore integer value from stack
          POP     R6              ;restore next stack value to R6
          POP     R4              ;restore next stack value to R4
          BNZ     LOOP_N          ;branch for next sample
          PUSH    R4              ;restore R4 into stack
          SUBI    1,R6            ;decrement repetition counter
          LDI     R6,R1           ;load R6 into R1
          PUSH    R6              ;save R6 into stack
          AND     1,R1            ;logical AND of 1 & R1
          BNZ     LOOP_F1         ;go to 1k Hz FAIL loop
          POP     R6              ;restore R6 stack
          BNZD    LOOP_F2         ;branch to 2 kHz FAIL signal
          PUSH    R6              ;save R6 into stack
          NOP                     ;no operation
          NOP                     ;no operation
          POP     R6              ;restore R6 from stack
          POP     R4              ;restore R4 from stack
          BZ      LOOP_N          ;get next sample
          .end                    ;end
```

FIGURE 8.6 *(continued)*

The pseudorandom noise generator program PRNOISE.ASM, or PRNOISEI.ASM in Chapter 3, produces a 1 or 0 (before scaling), and determines the frequency of the sinusoid to be generated. The scheme is to associate a 1 with an acceptable device and a 0 with a defective one. When the noise generator output is a 1, a 4-kHz sinusoid is generated, and when the noise generator output is a 0, a 2-kHz sinusoid followed by a 1-kHz sinusoid are generated.

The coefficients A and C ($B = -1$) in the recursive difference equation are calculated for a sampling frequency of 10 kHz and are set in the program.

The random noise generator produces the following sequence: 1, 1, 1, 1, 0, 1, 0, ... This causes the alarm program to generate the following sequence of tones with the frequencies: 4 kHz, 4 kHz, 4 kHz, 4 kHz (due to the first four values of 1's), followed by 2 kHz, 1kHz, 2 kHz, 1 kHz (due to the fifth value of 0), followed by 4 kHz (due to the sixth value of 1), followed by 2 kHz, 1kHz, 2kHz, 1kHz (due to the seventh value of 0), and so on. Four "fail" signals are generated for each noise sample of zero, because TIME_R = 4 in the program represents the number of times the fail signal is generated. A value of TIME_R = 1 causes the generation of the 2-kHz fail signal, not the 1-kHz signal, when a noise sample of zero is produced.

The amplitude of the 4-kHz "pass" signal is smaller (less pronounced) than the amplitude of the 2-kHz and 1-kHz "fail" signals.

Larger values for TIME_F and TIME_P would increase the duration of the generated tones, making it easier to hear the different pitches.

Run this program and verify these results.

8.4 EXTERNAL INTERRUPT FOR CONTROL

This project uses a hardware or external interrupt to control the amplitude of a sinewave. The sinewave generator using 4 points was introduced in Chapter 3 in the Experiment section. It is extended so that the amplitude of the generated sinewave is controlled using external interrupt.

The TMS320C31 supports a non-maskable external RESET signal, internal interrupts and four external maskable interrupts. The external interrupts are level-triggered using the status register for triggering. For a CPU interrupt to occur, the global interrupt enable GIE bit in the status register must be set and the interrupt flag IF register must also be set for interrupt enabled.

Figure 8.7 shows the external interrupt circuit. It includes a switch and a one-shot multivibrator chip. The two 32-pin connectors JP2 and JP6 along the edge of the DSK board provide the interrupt pin INT3, VCC, and ground.

Figure 8.8 shows a listing of the program EISINE.C which implements the amplitude control through hardware or external interrupt. Three interrupt functions are utilized. The serial port interrupt 05 generates the sinewave using a table look-up with 4 points. The serial port interrupt is used to determine the amount of time that has passed since an external interrupt occured, and turns the external interrupt back on after a desired delay count is reached.

The hardware external interrupt 04 increments the amplitude of the sinewave. Each time the switch is pressed, an external interrupt occurs, and the amplitude of the sinewave is incremented by 10%, through 10 levels. Then, the amplitude becomes zero.

Interrupt 10 specifies timer 1 interrupt. It provides deglitching, instead of using logic to prevent additional interrupts from occurring within a desired interval. Otherwise, several interrupts would occur. Switch bounce is accomplished by turning off the external interrupt for a period of time after it is first detected.

The period value of 0x30D4 = 12,500 corresponds to a rate of

$$\text{rate} = 12.5 \text{ MHz}/(2 \times \text{period}) = 500 \text{ Hz}$$

for a sample period of 2 ms. A delay count is incremented every 2 ms. When this delay count reaches 500, the external interrupt is turned back on. As a result, only one external interrupt can be acknowledged during an interval of (500 $\times$ 2 ms) = 1 s.

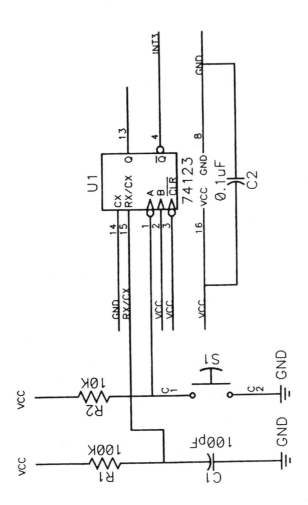

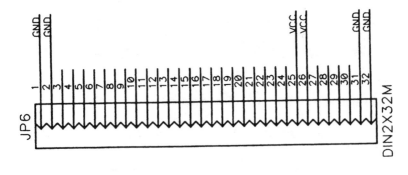

FIGURE 8.7 External interrupt circuit.

240

```
/*EISINE.C - SINE WITH 4 POINTS USING EXTERNAL INTERRUPT  */
#include "aiccomc.c"                /*AIC comm routines      */
int AICSEC[4] = {0x162C,0x1,0x4892,0x67}; /*AIC data        */
int data_out, loop = 0;   /*declare global variables         */
int sin_table[4] = {0,1000,0,-1000}; /*values for sinewave*/
int ampt_ctrl = 10;
int delay_cnt = 0;

void c_int10()
{
  if (delay_cnt < 500)
     delay_cnt++;
  else if (delay_cnt == 500)
  {
     asm("     AND      0FFFFFFF7h,IF");    /*set INT3 = 0 */
     asm("     OR       00000008h,IE");    /*set EINT3 = 1*/
  }
  else asm("    OR       00000008h,IE");
}


void c_int04()
{
   asm("     AND     0FFFFFFF7h,IE");       /*set EINT3 = 0*/
   delay_cnt = 0;
   if (ampt_ctrl < 10)
      ampt_ctrl++;
   else ampt_ctrl = 0;
}


void c_int05()
{
 int out;
 out = sin_table[loop] * ampt_ctrl * 0.1;
 PBASE[0x48] = out  << 2;   /*output value from sine table*/
 if (loop < 3) ++loop;      /*increment loop counter < 3   */
 else loop = 0;             /*reset loop counter           */
}

main()
{
```

(continued on next page)

FIGURE 8.8 External interrupt program for amplitude control of sinewave (EISINE.C).

```
PBASE[0x38] = 0x000030D4;  /*set timer 1 period        */
PBASE[0x30] = 0x000003C1;  /*set timer 1 control register*/
asm("    OR      00000208h,IE"); /*enable EINT3 & ETINT1*/
AICSET_I();                      /*configure AIC        */
for (;;);                        /*wait for interrupt   */
}
```

FIGURE 8.8   *(continued)*

Another version of this program is on disk as SINE4INT.ASM with only two interrupt functions: the hardware and the serial interrupts. It uses the sample rate of the serial port to achieve the desired delay for deglitching as was accomplished with interrupt 10.

8.5 MISCELLANEOUS APPLICATIONS AND PROJECTS

This section briefly discusses a number of projects that have been implemented by a number of students and described in [1–6]. These projects are based on the TMS320C30 EVM and can be extended for the DSK.

1. Acoustic Direction Tracker

This project discusses an acoustic signal tracker capable of tracking an audio source radiating a signal. It uses two microphones to capture the signal. From the delay associated with the signal reaching one of the microphones before the other, a relative angle where the source is located can be determined. The direction of the signal is displayed on the PC monitor and continuously updated.

A radiated signal at a distance from its source can be considered to have a plane wavefront, as shown in Figure 8.9. This allows the use of equally spaced sensors in a line to ascertain the angle at which the signal is radiating. While many microphones can be used as acoustical sensors, only two are used in this project. Since one microphone is closer to the source than the other, the signal received by the more-distant microphone is delayed in time. This time-shift corresponds to the angle where the source is located and the relative distance between the microphones and the source.

The angle $c = \arcsin(a/b)$, where the distance a is the product of the speed of sound and the time delay (phase/frequency). A simulation of the source-microphone relationship is obtained as shown in Figure 8.10 with the program SIM2.C (on the accompanying disk). The lower-left window in Figure 8.10 indicates the relative position of the source and the two microphones, represented, respectively, by the circle on the lower-left side and the small vertical line on the middle-right side. Press ALT-X to increase the distance between the two microphones.

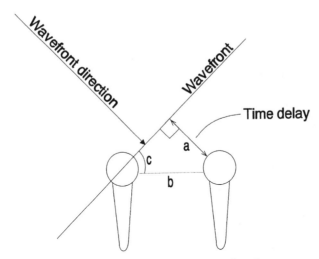

FIGURE 8.9 Signal reception with two microphones.

From Figure 8.10, one of the signals (gray) leads the other signal (blue). This indicates that the source is closer to the bottom microphone. Use the up-arrow key (from the keypad) to place the source position in line or in the middle of the two microphones. This represents that the two signals are now in phase. Use the up-arrow key to place the source above the microphones and verify that the signal that was leading before is now lagging.

Figure 8.11 shows a block diagram of the acoustic signal tracker. Two 128-point arrays of data are obtained, cross-correlating the first signal with the second one and then the second signal with the first one. The resulting cross-correlated data is decomposed into two halves, each transformed using the 128-point real-valued FFT function described in Chapter 6. The resulting phase is the phase difference of the two signals. Since two inputs are required, the input/output system described in Appendix D may be useful for this project.

This project was implemented with the TMS320C30-based EVM [3]. An external AIC board, connected through a second serial port available on the EVM provides the required second input [1]. The alternate use of the primary and auxiliary inputs on the same AIC was found to be inefficient. The acoustic tracker algorithms were verified with MATLAB, described in Appendix B, before a real-time implementation. This project was tested by positioning a speaker a few feet from the two microphones separated by one foot. The speaker received a 712-Hz signal from a function generator. The results are displayed on the PC monitor: a plot representing the track of the source-speaker over time as the speaker is slowly shifted from one side of the microphones to the other side. The PC monitor also displays a plot of both the cross-correlation and the magnitude of the cross-correlation of the two microphone signals.

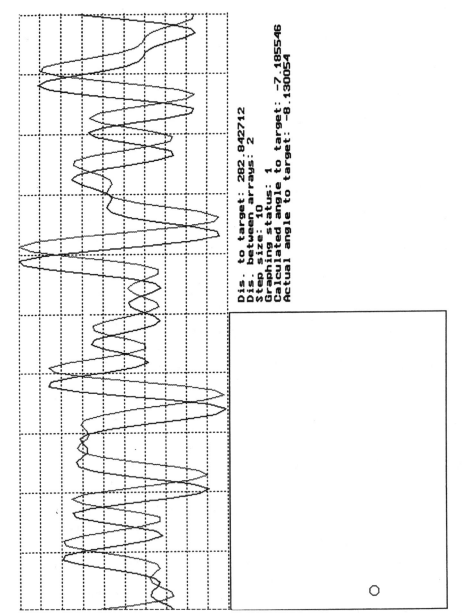

Dis. to target: 282.842712
Dis. between arrays: 2
Step size: 10
Graphing status: 1
Calculated angle to target: -7.185546
Actual angle to target: -8.130054

FIGURE 8.10 Plot of two simulated signals received by the microphones.

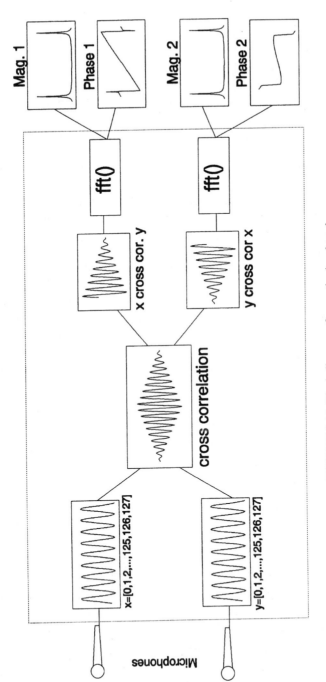

FIGURE 8.11 Block diagram of acoustic signal tracker.

2. Harmonic Analyzer

In a power system, the supplied voltage and the resulting load currents can be quite distorted. This is referred to as harmonic distortion. A large source of harmonic distortion is an AC-to-DC converter (rectifier) used in power supplies, etc. Another source of harmonic distortion is the switching-mode power supply used in microprocessor-based electronic equiment. While this type of power supply is more efficient and less expensive than the bulky power supplies, it is highly nonlinear and can be a major source of harmonic distortion and noise [18]. Other nonlinear sources of harmonics include arc furnaces, magnetic-saturation transformers, and fluorescent lights. As the use of nonlinear load becomes more widespread on the power-distribution system, the consequences of the resulting voltage and current distortions are becoming more significant. Harmonic distortion can result in overheating conductors, derating of transformers, generators, and motors, and noise and resonance problems in electrical distribution and communication systems. The need for real-time on-site data acquisition and analysis can be used by the electric utility industry to enable accurate measurements of harmonic distortion [3]. The harmonic analyzer could be installed between the electric utility and the customer or within the customer's facility to determine which equipment is producing the harmonic problems.

Figure 8.12 shows a block diagram of the major system components of the harmonic analyzer. Potentially dangerous voltages must be isolated from the digital electronic by a clamp-on current transformer (CT) and a potential trans-

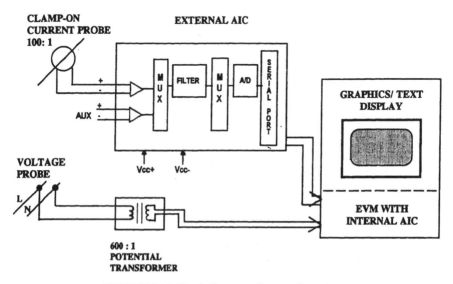

FIGURE 8.12 Block diagram of harmonic analyzer.

former (PT). The input voltage is the input to the AIC on board the EVM and the input current is a second input to an external AIC board [3]. The signal processing flow is illustrated in Figure 8.13. Results of the voltage and current waveforms are displayed on the PC monitor. A 100-watt incandescent light bulb as a linear load shows that minimal (if any) harmonics are generated. In contrast, the current waveform of a 15-watt fluorescent light bulb is not sinusoidal, and harmonics are generated, as displayed from the frequency plot of the current. A 512-point FFT was performed using the real-valued FFT function described in Chapter 6.

This project requires two inputs and can be extended for the DSK. The I/O board described in Appendix D provides for two inputs.

Possible enhancements to this project include seven in lieu of two channels, which would allow for the measurement of both voltage and current on three phases, and the neutral.

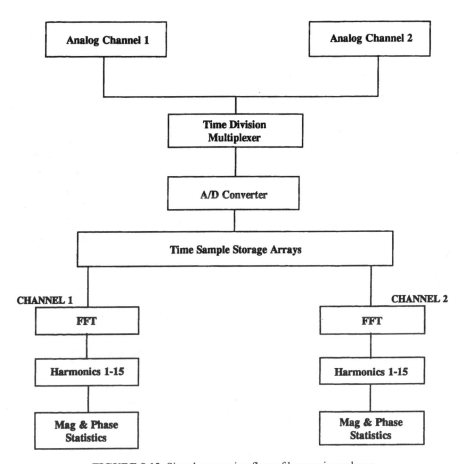

FIGURE 8.13 Signal processing flow of harmonic analyzer.

3. Speech Processing for Identification

Biometrics is a technology used for the verification or recognition of an individual, and employs methods for automated identification [2,19]. These methods include techniques for identifying an individual using physiological or behavioral characteristics. Physiological characteristics make use of the individual's hand, face, eye, and fingerprint. Behavioral characteristics such as voice and signature may vary from time to time, but are in general less costly to implement. This project focuses on speech as a means of verification in which an individual's identification is either accepted or rejected.

Speech information is primarily conveyed by the short time spectrum, the spectral information contained within a small time period. A direct approach is to uniformly chop a word into segments, with the idea that a subset of the words is enough to be recognized using match filter techniques. Recognition requires the comparison of stored characteristics with the characteristics presented.

This project is tested by having an individual speak a word, which is digitized, processed, and compared with a previously stored pattern. A PC host program executes a cross-correlation algorithm and a TMS320-based EVM program implements a 512-point real-valued FFT algorithm. A Hanning window function with 50% overlap allows the frequency data to be continuous between segments. Figure 8.14 displays on the PC monitor the speech data for identification. The upper plot (RECORD1) represents the spoken word GOD and the middle plot (RECORD2) represents the stored template of the word GOD. The lower plot (COMPARE) compares or cross-correlates the two previous sets of data. Cross-correlation is achieved by multiplying each element in the first segment of RECORD1 data by each element of the three segments of RECORD2. This process is repeated for the other two segments of RECORD1 to provide three arrays of (3 × 512) points and indicate like-frequency components between the data of RECORD1 and RECORD2. Figure 8.14 designates a match. The level of a match can be adjusted so that if the correlation result reaches a predetermined level, a match is specified.

4. FFT-Based Security System

The scheme in this project is to simulate and detect the presence of an intruder in each of four monitored zones. A sensor circuit, designed and built with infrared modules, provides a composite waveform signal that contains frequency information related to the status of a security system. The FFT of this signal determines which zones are detecting the presence of an intruder. The sensor circuit includes infrared detection modules, frequency division counters, logic gates, and a summing amplifier [3]. The infrared detection modules consist of an infrared diode and transistor pair contained in a single package. The module's transistor detects the infrared wave emitted by the diode if the transmission path between the diode–transistor pair is unobstructed. If an instruder is present,

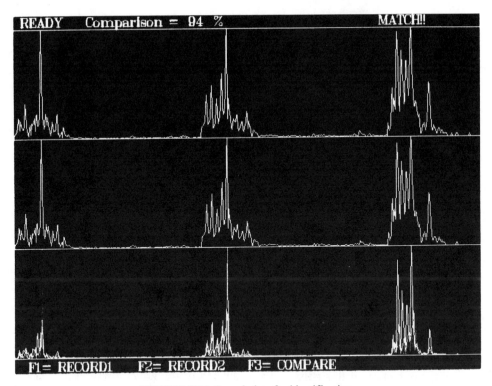

FIGURE 8.14 Speech data for identification.

the transmission path is blocked and the infrared wave is not detected. The comparator's output level is a high or a low, based on the transistor's collector-to-ground voltage compared to a known reference voltage, which corresponds to the presence or the absence of an intruder.

The frequency-division counters divide down a 4-MHz oscillator into four distinct square waves with the following frequencies: 1, 2, 2.5, and 4 kHz. Each frequency represents one of four zones that the security system monitors. The summing amplifier mixes and amplifies the frequencies at its input into one composite signal, which becomes the AIC input. If no intruder is detected, all four zone frequencies are passed to the summing amplifier. If an intruder is detected, the zone frequency associated with the detector sensing the intruder is eliminated from the composite signal, or prevented from reaching the output of the detector circuit.

The real-valued FFT function performs a 128-point FFT on the composite signal from the infrared circuit. From the resulting magnitude response, a PC host program determines the frequencies present. For example, if the transmission paths of the 1-kHz and the 4-kHz sensors are blocked, an Alert message flashes on the PC monitor in the stations associated with these two frequencies.

Projects Implemented With the TMS320C30 EVM Described in Reference 1

The following 11 application projects were implemented with the TMS320C30-based EVM and are described in Reference 1.

1. *Parametric Equalizer.* A parametric equalizer is basically an audio-frequency shaper similar to the more popular graphic equalizer. The equalizer implemented consists of 16 different sets of filters, each with two passbands that are used to increase or decrease the amplitude of a selected range of frequencies. Each band is controlled independently of the other, with amplitude gain or attenuation and adjustable center frequency and bandwidth. The first project with the eight sets of FIR filters, described in this chapter, evolved from the Parametric Equalizer project.

2. *Adaptive Notch Filter Using TMS320C30 Code.* Three types of filters are implemented to illustrate the reduction of a 60-Hz artifact in an electrocardiogram (ECG) signal: a 60-Hz notch filter and two adaptive filters, one with two weights, as discussed in Chapter 7. Sinusoidal 60-Hz interference is a frequent problem in electrocardiographic (ECG) monitoring, creating baseline artifacts that can obscure the true ECG waveform and hinder diagnostic interpretation of the ECG. The ECG is a representation of the electrical impulses that are associated with cardiac muscle contraction and relaxation. Figure 8.15 is a display of an ECG monitor initially showing the 60-Hz sinusoidal noise before adaptation, and the reduction of this noise in the latter part of the waveform after adaptation. The negative spikes represent a 2-Hz ECG signal.

3. *Adaptive Filter for Noise Cancellation Using C Code.* This project is a real-time implementation of the adaptive predictor structure described in Chapter 7. To obtain the delay, a table-lookup scheme is used. The input consists of a 2-kHz desired signal with added random noise and the output is shown to converge to the 2-kHz input signal.

4. *Swept Frequency Response.* The digital oscillator discussed in Chapter 5 is extended in this project to include a frequency-sweep feature in C code. The project is tested by using the output-swept frequency from the EVM as input to a second-order bandpass R-L-C analog filter and the analog filter's output as input to the EVM. The PC monitor displays the frequency response of the second-order analog filter.

5. *Multirate Filter.* This project implements on the EVM the multirate filter described in Section 8.2.

6. *Introduction to Image Processing: Video Line Rate Analysis.* This project analyzes a video signal at the horizontal (line) rate using C code. Interactive algorithms commonly used in image processing for filtering, averaging, and edge enhancement are utilized for this analysis. The source of the video signal is a charge coupled device (CCD) camera as input to a module designed and built for this project. This module includes flip-flops, logic gates, an ADC, and a 9.8-MHz clock [1,6]. Figure 8.16 (a) shows a display on the PC monitor screen of

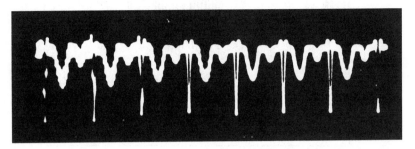

FIGURE 8.15 Adaptive filtering of 60-Hz noise displayed on ECG monitor.

one horizontal video line signal, with a 500-kHz sixth-order IIR lowpass filter *on* in Figure 8.16 (b), and with edge enhancement algorithm *on* in Figure 8.16 (c). Note that the function key F4 turns *on* a 3-MHz lowpass filter, passing higher-frequency components.

7. *PID Controller.* This project implements a digital speed-control system for a servomotor using a well-known proportional, integral, and derivative (PID) control algorithm in C code. The system includes a DC motor used as a servomotor and a tachgenerator to translate speed into voltage. The structure of the PID controller, the block diagram, and the driver circuit of the control system are described in References 1 and 4. The user is prompted to enter the proportional, integral, and derivative gain constants. When the desired speed is obtained, the output of the tachgenerator remains constant until the system is disturbed. If friction is applied, the algorithm increases an output control voltage, so that the desired speed is maintained. Figure 8.17 shows the effects of PID gain constants on the motor speed, with proportional, integral, and derivative gain constants of 10, 0.01, and 50, respectively.

8. *Wireguided Submersible.* A wireguided submersible system is controlled via a wire, enabling the submersible to be steered in the water. The control sends signals of different frequencies to the submersible. The submersible is maneuvered based on the specific frequency of the received signal. This project, implemented in C, transforms a received signal from a function generator using a 512-point real-valued FFT. The PC monitor screen displays the simulated submersible course, with dots on the screen "moving" to the right or to the left depending on whether the received signal is 1 or 2 kHz, respectively. Another indicator that represents the depth moves up or down depending on whether the received signal is 3 or 4 kHz, respectively. Outside a frequency range set in the program, the dots "move" in a straight line [1,4].

This project can be extended in many ways. A microphone can be used as the input control. The submersible can filter out any noise received with a control signal. Multiple frequencies can be sent down the wire to the submersible, which would then aknowledge the received control signal and specify the re-

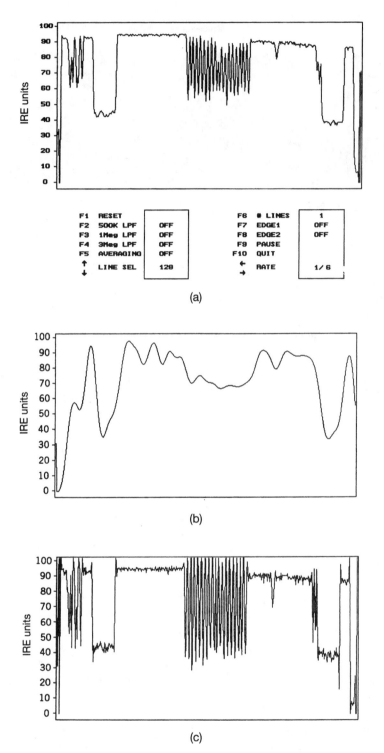

FIGURE 8.16 PC display of one horizontal video line signal (a) with no image processing algorithm *on*; (b) with 500-kHz lowpass filter *on*; (c) with edge enhancement algorithm *on*.

sponse or the course being followed. Also the duration of the control signal can specify the course to be taken.

9. *Frequency Shift Using Modulation.* In this project, a 1-kHz input is used to generate a 1.001, 2, 3, or 4 kHz modulated signal, implemented in C code [1,4]. The modulated frequency signal is $y(n) = x\cos(2\pi nf/F_s)$, where x is the 1-kHz input signal used as a carrier frequency signal and f is the desired shifted frequency. The modulated signal $y(n)$ becomes the input to an IIR filter selected among four sixth-order IIR bandpass filters, centered at 1, 2, 3, and 4 kHz, respectively.

10. *Four-Channel Multiplexer For Fast Data Acquisition.* A four-channel multiplexer module is designed and built for this project, implemented in C code [1,4]. It includes an 8-bit flash ADC, a FIFO, a MUX, and a 2-MHz crystal oscillator. An input sinusoid is acquired through one of the four channels selected with the function keys F1 through F4. A 128-point real-valued FFT on the input signal is displayed in real-time on the PC monitor screen. A 20-MHz oscillator was also used, but it produced a noisier signal.

11. *Neural Network for Signal Recognition.* The FFT of a signal becomes the input to a neural network, which is trained to recognize this input signal using a back-propagation learning rule [1,4,20,21], implemented in C code. Many different rules are available for training a neural network, and the back-propagation is one of the most widely used for a wide range of applications. Given a set

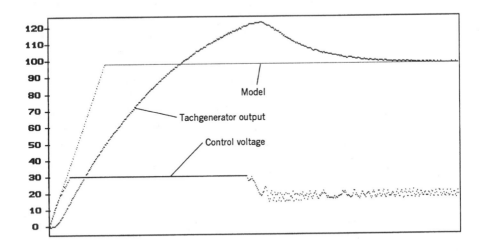

```
PGAIN = 10
IGAIN = 0.01
DGAIN = 50
```

FIGURE 8.17 Effects of PID gain constants on motor speed.

of inputs, the network is trained to give a desired response. If the network gives the wrong answer, then the network is corrected by adjusting its parameters (weights) so that the error is reduced. During this correction process, one starts with the output nodes and propagation is backward to the input nodes.

Reference [7] describes 27 projects associated with the fixed-point processor, and can provide a good source of ideas for projects.

REFERENCES

1. R. Chassaing, *Digital Signal Processing with C and the TMS320C30,* Wiley, New York, 1992.

2. R. Chassaing et al., "Student Projects on Digital Signal Processing with the TMS320C30," in *Proceedings of the 1995 ASEE Annual Conference,* June 1995.

3. R. Chassaing et al., "Digital Signal Processing with C and the TMS320C30: Senior Projects," in *Proceedings of the Third Annual TMS320 Educators Conference,* Texas Instruments, Inc., Dallas, TX, 1993.

4. R. Chassaing et al., "Student Projects on Applications in Digital Signal Processing with C and the TMS320C30," in *Proceedings of the Second Annual TMS320 Educators Conference,* Texas Instruments Inc., Dallas, TX, 1992.

5. R. Chassaing, "TMS320 in a Digital Signal Processing Lab," in *Proceedings of the TMS320 Educators Conference,* Texas Instruments Inc., Dallas, TX, 1991.

6. B. Bitler and R. Chassaing, "Video Line Rate Processing with the TMS320C30," in *Proceedings of the 1992 International Conference on Signal Processing Applications and Technology (ICSPAT),* 1992.

7. R. Chassaing and D. W. Horning, *Digital Signal Processing with the TMS320C25,* Wiley, 1990.

8. P. Papamichalis ed., *Digital Signal Processing Applications with the TMS320 Family— Theory, Algorithms, and Implementations,* Vols. 2 & 3, Texas Instruments, Inc., Dallas, TX, 1989 and 1990.

9. K. S. Lin ed., *Digital Signal Processing Applications with the TMS320 Family—Theory, Algorithms, and Implementations,* Vol. 1, Texas Instruments Inc., Dallas, TX, 1987.

10. R. E. Crochiere and L. R. Rabiner, *Multirate Digital Signal Processing,* Prentice-Hall, Englewood Cliffs, NJ, 1983.

11. R. W. Schafer and L. R. Rabiner, "A digital signal processing approach to interpolation," *Proceedings of the IEEE,* **61,** 692–702 (1973).

12. R. E. Crochiere and L. R. Rabiner, "Optimum FIR Digital Filter Implementations for Decimation, Interpolation, and Narrow-Band Filtering," *IEEE Trans. on Acoustics, Speech, and Signal Processing,* **ASSP-23,** 444–456 (1975).

13. R. E. Crochiere and L. R. Rabiner, "Further Considerations in the Design of Decimators and Interpolators," *IEEE Trans. on Acoustics, Speech, and Signal Processing,* **ASSP-24,** 296–311 (1976).

14. M. G. Bellanger, J. L. Daguet, and G. P. Lepagnol, "Interpolation, Extrapolation, and Reduction of Computation Speed in Digital Filters," *IEEE Trans. on Acoustics, Speech, and Signal Processing,* **ASSP-22,** 231–235 (1974).

15. R. Chassaing, P. Martin, and R. Thayer, "Multirate Filtering Using the TMS320C30 Floating-Point Digital Signal Processor," in *Proceedings of the 1991 ASEE annual Conference,* June 1991.

16. R. Chassaing, "Digital Broadband Noise Synthesis by Multirate Filtering Using the TMS320C25," in *Proceedings of the 1988 ASEE Annual Conference,* Vol. 1, June 1988.

17. R. Chassaing, W. A. Peterson, and D. W. Horning, "A TMS320C25-Based Multirate Filter," *IEEE Micro,* October 1990, pp. 54–62.

18. A. Freund ed., "Nonlinear Loads Mean Trouble," *EC & M,* March 1988.

19. B. Miller, "Biometric Identification," *IEEE Spectrum,* Feb. 1994.

20. B. Widrow and R. Winter, "Neural Nets for Adaptive Filtering and Adaptive Pattern Recognition," *Computer Magazine,* Computer Society of the IEEE, March 1988, pp. 25–39.

21. D. E. Rumelhart, J. L. McClelland, and the PDP Research Group, *Parallel Distributed Processing: Explorations in the Microstructure of Cognition,* Vol. 1, MIT Press, Cambridge, MA, 1986.

A

TMS320C3x Instruction Set and Registers

The TMS320C3x contains a rich set of instructions as well as a number of useful registers used for I/O, interrupt, etc. [1]

A.1 TMS320C3x INSTRUCTION SET

Tables A.1–A.5 show a summary of the TMS320C3x instruction set. They contain load and store instructions, two- and three-operand instructions, as well as program control and parallel instructions.

A.2 TMS320C3x REGISTERS

The TMS320C31 contains a number of registers that are very useful during program development.

TABLE A.1 Load and store instructions (reprinted by permission of Texas Instruments, Inc.)

Instruction	Description	Instruction	Description
LDE	Load floating-point exponent	POP	Pop integer from stack
LDF	Load floating-point value	POPF	Pop floating-point value from stack
LDFcond	Load floating-point value conditionally	PUSH	Push integer on stack
LDI stack	Load integer	PUSHF	Push floating-point value on
LDIcond	Load integer conditionally	STF	Store floating-point value
LDM	Load floating-point mantissa	STI	Store integer
LDP	Load data page pointer		

TABLE A.2 Two-Operand instructions (reprinted by permission of Texas Instruments, Inc.)

Instruction	Description	Instruction	Description
ABSF	Absolute value of a floating-point number	NORM	Normalize floating-point value
ABSI	Absolute value of an integer	NOT	Bitwise-logical complement
ADDC†	Add integers with carry	OR†	Bitwise-logical OR
ADDF†	Add floating-point values	RND	Round floating-point value
ADDI†	Add integers	ROL	Rotate left
AND†	Bitwise-logical AND	ROLC	Rotate left through carry
ANDN†	Bitwise-logical AND with complement	ROR	Rotate right
ASH†	Arithmetic shift	RORC	Rotate right through carry
CMPF†	Compare floating-point values	SUBB†	Subtract integers with borrow
CMPI†	Compare integers	SUBC	Subtract integers conditionally
FIX	Convert floating-point value to integer	SUBF†	Subtract floating-point values
FLOAT	Convert integer to floating-point value	SUBI†	Subtract integer
LSH†	Logical shift	SUBRB	Subtract reverse integer with borrow
MPYF†	Multiply floating-point values	SUBRF	Subtract reverse floating-point value
MPYI†	Multiply integers	SUBRI	Subtract reverse integer
NEGB	Negate integer with borrow	TSTB†	Test bit fields
NEGF	Negate floating-point value	XOR†	Bitwise-exclusive OR
NEGI	Negate integer		

† 2- and 3-operand versions

TABLE A.3 Three-Operand instructions (reprinted by permission of Texas Instruments, Inc.)

Instruction	Description	Instruction	Description
ADDC3	Add with carry	MPYF3	Multiply floating-point values
ADDF3	Add floating-point values	MPYI3	Multiply integers
ADDI3	Add integers	OR3	Bitwise-logical OR
AND3	Bitwise-logical AND	SUBB3	Subtract integers with borrow
ANDN3	Bitwise-logical AND with complement	SUBF3	Subtract floating-point values
ASH3	Arithmetic shift	SUBI3	Subtract integers
CMPF3	Compare floating-point values	TSTB3	Test bit fields
CMPI3	Compare integers	XOR3	Bitwise-exclusive OR
LSH3	Logical shift		

TABLE A.4 Program control instructions (reprinted by permission of Texas Instruments, Inc.)

Instruction	Description	Instruction	Description
B*cond*	Branch conditionally (standard)	IDLE	Idle until interrupt
B*cond*D	Branch conditionally (delayed)	NOP	No operation
BR	Branch unconditionally (standard)	RETI*cond*	Return from interrupt conditionally
BRD	Branch unconditionally (delayed)	RETS*cond*	Return from subroutine conditionally
CALL	Call subroutine	RPTB	Repeat block of instructions
CALL*cond*	Call subroutine conditionally	RPTS	Repeat single instruction
DB*cond*	Decrement and branch conditionally (standard)	SWI	Software interrupt
DB*cond*D	Decrement and branch conditionally (delayed)	TRAP*cond*	Trap conditionally
IACK	Interrupt acknowledge		

1. The status register (ST) format, shown in Figure A.1, provides information about the state of the CPU. The condition flags of the ST register are set based on resulting operations.

2. The interrupt enable (IE) register format is shown in Figure A.2. A 1 (or a 0) enables (or disables) an interrupt.

3. The memory-mapped interrupt locations are shown in Figure A.3. For example, XINT0 is mapped to the memory address 0x809C05.

4. The addresses of the peripheral bus mapped registers are shown in Figure A.4. Note the addresses of the timer registers used for interrupt and the serial port registers for communicating with the on-board AIC. The timer global control register format and the serial port global control register format are shown in Figures A.5 and A.6, respectively. The mode of the timer is specified by the timer global control register, and the sampling frequency is determined by the timer period register. The timer counter register is for incrementing the count from zero to the period register value.

5. The interrupt flag (IF) register format is shown in Figure A.7. An interrupt is set with a 1 or cleared with a 0 in a flag register bit.

6. The I/O flag register format is shown in Figure A.8. It controls the function of the external pins XF0 and XF1 for I/O. This register is set to 0 on reset.

TABLE A.5 Parallel instructions (reprinted by permission of Texas Instruments, Inc.)

Mnemonic	Description
ABSF ‖ STF	Absolute value of a floating-point number and store floating-point value
ABSI ‖ STI	Absolute value of an integer and store integer
ADDF3 ‖ STF	Add floating-point values and store floating-point value
ADDI3 ‖ STI	Add integers and store integer
AND3 ‖ STI	Bitwise-logical AND and store integer
ASH3 ‖ STI	Arithmetic shift and store integer
FIX ‖ STI	Convert floating-point to integer and store integer
FLOAT ‖ STF	Convert integer to floating-point value and store floating-point value
LDF ‖ STF	Load floating-point value and store floating-point value
LDI ‖ STI	Load integer and store integer
LSH3 ‖ STI	Logical shift and store integer
MPYF3 ‖ STF	Multiply floating-point values and store floating-point value
MPYI3 ‖ STI	Multiply integer and store integer
NEGF ‖ STF	Negate floating-point value and store floating-point value
NEGI ‖ STI	Negate integer and store integer
NOT ‖ STI	Complement value and store integer
OR3 ‖ STI	Bitwise-logical OR value and store integer
STF ‖ STF	Store floating-point values
STI ‖ STI	Store integers
SUBF3 ‖ STF	Subtract floating-point value and store floating-point value
SUBI3 ‖ STI	Subtract integer and store integer
XOR3 ‖ STI	Bitwise-exclusive OR values and store integer
LDF ‖ LDF	Load floating-point value
LDI ‖ LDI	Load integer
MPYF3 ‖ ADDF3	Multiply and add floating-point value
MPYF3 ‖ SUBF3	Multiply and subtract floating-point value
MPYI3 ‖ ADDI3	Multiply and add integer
MPYI3 ‖ SUBI3	Multiply and subtract integer

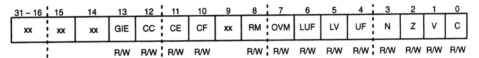

Notes: 1) xx = reserved bit, read as 0
2) R = read, W = write

FIGURE A.1 Status (ST) register format (reprinted by permission of Texas Instruments, Inc.).

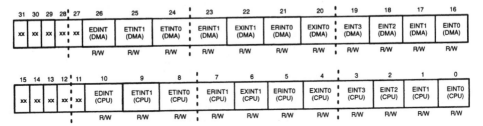

FIGURE A.2 Interrupt enable (IE) register format (reprinted by permission of Texas Instruments, Inc.).

Address	Content
809FC1h	INT0
809FC2h	INT1
809FC3h	INT2
809FC4h	INT3
809FC5h	XINT0
809FC6h	RINT0
809FC7h	XINT1 (reserved)
809FC8h	RINT1 (reserved)
809FC9h	TINT0
809FCAh	TINT1
809FCBh	DINT
809FCCh – 809FDFh	Reserved
809FE0h	TRAP 0
809FE1h	TRAP 1
	• • •
809FFBh	TRAP 27
809FFCh	TRAP 28 (reserved)
809FFDh	TRAP 29 (reserved)
809FFEh	TRAP 30 (reserved)
809FFFh	TRAP 31 (reserved)

FIGURE A.3 Memory-mapped interrupt locations (reprinted by permission of Texas Instruments, Inc.).

808000h	DMA global control
808004h	DMA source address
808006h	DMA destination address
808008h	DMA transfer counter
808020h	Timer 0 global control
808024h	Timer 0 counter
808028h	Timer 0 period
808030h	Timer 1 global control
808034h	Timer 1 counter
808038h	Timer 1 period register
808040h	Serial port global control
808042h	FSX/DX/CLKX serial port control
808043h	FSR/DR/CLKR serial port control
808044h	Serial port R/X timer control
808045h	Serial port R/X timer counter
808046h	Serial port R/X timer period
808048h	Serial port data transmit
80804Ch	Serial port data receive
808064h	Primary-bus control

FIGURE A.4 Peripheral bus memory-mapped registers (reprinted by permission of Texas Instruments, Inc.).

31 16	15 12	11	10	9	8	7	6	5	4	3	2	1	0
xx	xx	TSTAT	INV	CLKSRC	C/$\overline{\text{P}}$	$\overline{\text{HLD}}$	GO	xx	xx	DATIN	DATOUT	$\overline{\text{I}}$/O	FUNC
		R	R/W	R/W	R/W	R/W	R/W			R	R/W	R/W	R/W

Notes: 1) R = read, W = write

2) xx = reserved bit, read as 0

FIGURE A.5 Timer global control register format (reprinted by permission of Texas Instruments, Inc.).

FIGURE A.6 Serial port global control register format

31	30	29	28	27	26	25	24	23	22	21	20	19	18	17	16
xx	xx	xx	xx	RRESET	XRESET	RINT	RTINT	XINT	XTINT	RLEN		XLEN		FSRP	FSXP
				R/W	R/W	R/W	R/W	R/W	R/W	R/W	R/W		R/W	R/W	R/W

15	14	13	12	11	10	9	8	7	6	5	4	3	2	1	0
DRP	DXP	CLKRP	CLKXP	RFSM	XFSM	RVAREN	XVAREN	RCLK SRCE	XCLK SRCE	HS	RSR FULL	XSR EMPTY	FSXOUT	XRDY	RRDY
R/W	R/W	R/W	R/W	R/W	R/W	R/W	R/W	R/W	R/W	R/W	R	R	R/W	R	R

Notes: 1) R = read, W = write
       2) xx = reserved bit, read as 0

FIGURE A.6 Serial port global control register format (reprinted by permission of Texas Instruments, Inc.).

31–16	15–12	11	10	9	8	7	6	5	4	3	2	1	0
xx	yy	yy	DINT	TINT1	TINT0	xx	xx	RINT0	XINT0	INT3	INT2	INT1	INT0
			R/W	R/W	R/W	R/W	R/W	R/W	R/W	R/W	R/W	R/W	R/W

Notes: 1) xx = reserved bit, read as 0
       2) yy = reserved bit, set to 0 at reset
       3) R = read, W = write

FIGURE A.7 Interrupt flag (IF) register format (reprinted by permission of Texas Instruments, Inc.).

31–16	15–12	11–8	7	6	5	4	3	2	1	0
xx	x	xx	INXF1	OUTXF1	I/OXF1	xx	INXF0	OUTXF0	I/OXF0	xx
			R	R/W	R/W		R	R/W	R/W	

Notes: 1) xx = reserved bit, read as 0
       2) R = read, W = write

FIGURE A.8 I/O flag (IOF) register format (reprinted by permission of Texas Instruments, Inc.).

REFERENCE

1. *TMS320C3x User's Guide,* Texas Instruments, Inc., Dallas, TX, 1997.

B

Support Tools

B.1 CODE EXPLORER DEBUGGER FROM GO DSP

The debugger Code Explorer from GO DSP, available from the web [1] is an abridged version of the popular debugger Code Composer, and supports the C31 DSK. It does not support executable COFF files (Code Composer does). Appropriate documentation is provided with Code Composer.

With Code Explorer, one can single-step through an executable DSK file, run with breakpoints, etc. A nice feature is that one can plot a set of data in both the time and the frequency domains as illustrated in the following example.

Example B.1 FIR Filter Using Code Explorer For Debugging and Plotting

To Load and Run

Access and run Code Explorer. Figure B.1 shows a window screen of the Code Explorer debugger. Proceed with the following:

1. Select File, Load Program, and select the file BP45SIMP.DSK (on the accompanying disk) to download into the debugger. This source file BP45SIMP.ASM, shown in Figure B.2, is similar to BP45SIM.ASM, discussed in Chapter 4, but uses 64 input data values (10,000, 0, . . .) and an output buffer of length 64 initialized to zero. The filter routine is executed 64 times, yielding a total of 64 output data samples, in lieu of 45 output data samples obtained with BP45SIM.ASM. The filter's length is 45 (with 45 coefficients).

2. Click on Run to execute and on Halt to stop execution.

3. To verify the output values, select View, then Memory. Enter OUTB for address and select 32-bit Signed Integer format. Verify that the output values starting at the address OUTB or 0x809C71 are –19, –27, . . . , –19, 0, 0, . . . , which are the impulse response coefficients scaled by 10,000 followed by 0, 0, ... Note that the output sequence is essentially padded (at the end) with 19

265

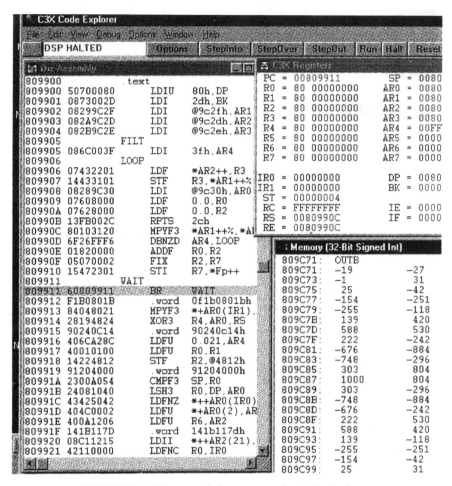

FIGURE B.1 Debugger window screen using Code Explorer.

zeros between 0x809C9E and 0x809CB0. These extra output sample points allow for a plot with a higher resolution.

To Create An Output File

1. Select File, then File I/O, and then File Output.

2. Select Add File, then type C:\dsktools\BP45SIMP.DAT to designate the path dsktools and BP45SIMP.DAT as the output file.

3. Select OUTB for address that is the starting output address, and select a length of 64.

4. Select Add Probepoint (Probepoint is not connected yet).

```
;BP45SIMP.ASM - FIR FILTER WITH 64 SAMPLE POINTS FOR SIMULATION
        .start   ".text",0x809900 ;where text begins
        .start   ".data",0x809C00 ;where data begins
        .include "BP45.COF"        ;include coefficients file
        .data                      ;data section
IN_ADDR  .word    INB              ;starting address for input
OUT_ADDR .word    OUTB             ;starting address for output
XB_ADDR  .word    XN+LENGTH-1      ;bottom address of circ buffer
HN_ADDR  .word    COEFF            ;starting addr of coefficients
NSAMPLE  .set     64               ;number of sample points


INB      .float   10000,0,0,0,0,0,0,0,0,0,0,0,0,0,0,0,0,0,0,0,0,0
         .float   0,0,0,0,0,0,0,0,0,0,0,0,0,0,0,0,0,0,0,0,0,0,0,0
         .float   0,0,0,0,0,0,0,0,0,0,0,0,0,0,0,0,0,0,0,0
OUTB     .float   0,0,0,0,0,0,0,0,0,0,0,0,0,0,0,0,0,0,0,0,0,0,0,0
         .float   0,0,0,0,0,0,0,0,0,0,0,0,0,0,0,0,0,0,0,0,0,0,0,0
         .float   0,0,0,0,0,0,0,0,0,0,0,0,0,0,0,0


         .brstart "XN_BUFF",64     ;align samples buffer
XN       .sect    "XN_BUFF"        ;section for input samples
         .loop    LENGTH           ;buffer size for samples
         .float   0.0              ;initialize samples to zero
         .endloop                  ;end of loop
         .entry   BEGIN            ;start of code


         .text                     ;text section
BEGIN    LDP      IN_ADDR          ;init to data page 128
         LDI      LENGTH,BK        ;size of circular buffer
         LDI      @XB_ADDR,AR1     ;last sample address ->AR1
         LDI      @IN_ADDR,AR2     ;input address       -->AR2
         LDI      @OUT_ADDR,AR3    ;output address      -->AR3
FILT     LDI      NSAMPLE-1,AR4    ;number of sample points-->AR4
LOOP     LDF      *AR2++,R3        ;input new sample
         STF      R3,*AR1++%       ;store newest sample
         LDI      @HN_ADDR,AR0     ;AR0 points to H(N-1)
         LDF      0,R0             ;init R0
         LDF      0,R2             ;init R2
         RPTS     LENGTH-1         ;repeat LENGTH-1 times
```

(continued on next page)

FIGURE B.2 FIR filter program with 64 samples for simulation (`BP45SIMP.ASM`).

```
        MPYF3    *AR0++,*AR1++%,R0 ;R0 = HN*XN
||      ADDF3    R0,R2,R2          ;accumulation in R2
        DBNZD    AR4,LOOP          ;delayed branch until AR4<0
        ADDF     R0,R2             ;last mult result accumulated
        FIX      R2,R7             ;convert float R2 to integer R7
        STI      R7,*AR3++         ;store into output buffer
WAIT    BR       WAIT              ;wait
        .end                       ;end
```

FIGURE B.2 *(continued)*

5. From the Probepoint screen, specify address OUTB, then select Con-nect to: FILEOUT:C:\..\BP45SIMP.DAT.

6. Select Add. This displays Probe Points:

$$\sqrt{}\ \ \text{OUTB} \rightarrow \text{File OUT:C:\..\BP45SIMP.DAT}$$

7. Select OK. Display shows that Probepoint is now connected (from the File Output Window screen). Then select OK. The output data file is now created as BP₄5SIMP.DAT, but it is not saved yet.

To Store/Save Output on Disk

Select File, then Store Data, and then BP45SIMP.DAT. Select Yes to replace existing file. Otherwise, the output file will contain only a heading. Select address OUTB and length 64. This will store the 64 output data values in the file BP45SIMP.DAT.

To Plot/Graph the Output

1. Select View, then Graph.

2. Specify OUTB for starting address, 64 for both the Buffer size and Display size, 10000 for Sampling Frequency, Line Graph, 32-bit Signed Integer, and Time Domain. Press OK to obtain a plot of the impulse response shown in Figure B.3.

3. Change Time Domain to Frequency Domain:FFT to get the frequency response plot shown in Figure B.4, which represents the characteristics of an FIR bandpass filter centered at $F_s/10$, as discussed in Chapter 4.

The 64 output sample points provide a graph with a better resolution than one with 45 points, as can be verified by repeating the above procedure with the program BP45SIM.DSK.

Repeat this example with the lowpass filter LP11SIM.ASM described in Chapter 4, and plot the output frequency response of this lowpass filter.

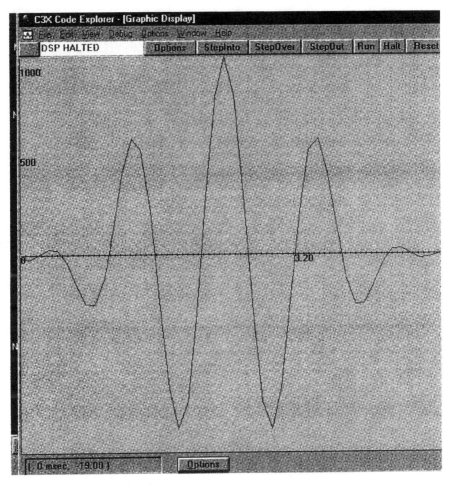

FIGURE B.3 Impulse response of FIR filter using Code Explorer.

B.2 VIRTUAL INSTRUMENT USING SHAREWARE UTILITY PACKAGE

Goldwave is a shareware utility software program that can turn a PC with a sound card into a virtual instrument. It can be downloaded from the web [2]. With Goldwave, one can create a function generator to generate different signals, such as sine and random noise. It can also be used as an oscilloscope or as a spectrum analyzer. To use and test it as both a signal generator and as a spectrum analyzer, proceed as follows.

   1. Access two copies of Goldwave running under Windows 95.

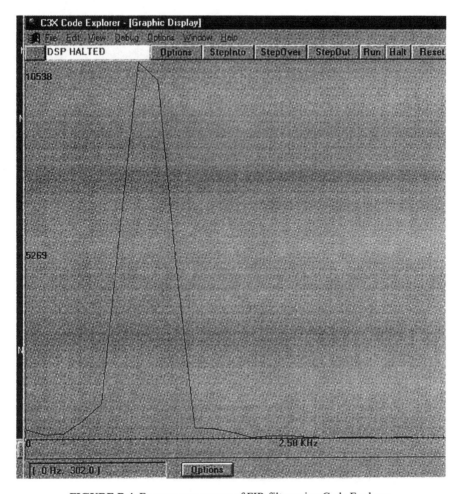

FIGURE B.4 Frequency response of FIR filter using Code Explorer.

2. Run a filter program, such as FIRNC.ASM from Chapter 4, on the DSK under DOS.

3. Access the available function $f(x)$ within one copy of Goldwave and create the random noise signal rand[1]. The output of the sound card becomes the DSK input running the FIR filter. Use the Device Control Setup with Goldwave to generate a continuous random noise input by "looping" back the noise signal a large number of times.

4. The output of the DSK becomes the input to the sound card. Access the second copy of Goldwave to use as a spectrum analyzer. The Device Control Setup allows you to obtain an oscilloscope or a spectrum analyzer.

5. Different sampling rates are supported. Figure B.5 shows the frequency response of a 41-coefficient FIR bandpass filter, centered at $F_s/4$. This is obtained by running `FIRNC.ASM` with the `BP41.COF` coefficient file included, and a sampling rate of 11.025 kHz chosen within Goldwave. An averaging of the waveform signal is not supported.

Figure B.6 shows a window screen with some segment of a speech waveform within Goldwave. This speech signal segment was recorded with Goldwave and stored as a `.wav` file. Effects such as echo and filtering can be readily observed. Lowpass, highpass, bandpass, and bandstop filters can be implemented with Goldwave and their effects on a signal illustrated.

Other shareware utility programs such as Cool Edit [3] or Spectrogram [4] can also be used as virtual spectrum analyzers.

B.3 FILTER DESIGN AND IMPLEMENTATION USING DIGIFILTER

DigiFilter, available from MultiDSP, is a filter design package that supports directly the C31 DSK [5]. It allows for the design and implementation of FIR and IIR filters in real-time on the DSK without any programming.

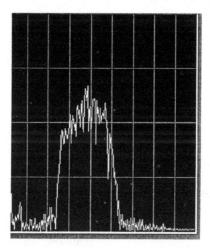

FIGURE B.5 Frequency response of 41-coefficient FIR bandpass filter using Goldwave.

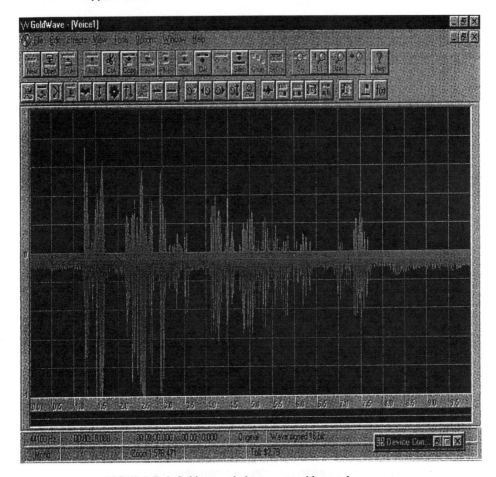

FIGURE B.6 Goldwave window screen with speech segment.

Example B.2 FIR Filter Design and Implementation Using DigiFilter

Figure B.7 shows a plot of the Log Magnitude response of a 61-coefficient FIR bandpass filter, centered at 2 kHz using the Kaiser window function. For a specific design, the user can select among several window functions, with the specification of the number of taps (coefficients) associated with each window (rectangular, Hamming, etc.). Impulse as well as step responses can also be obtained, as shown in Figure B.8. Note that an implementation with a Hamming window function would require 89 coefficients (Figure B.8).

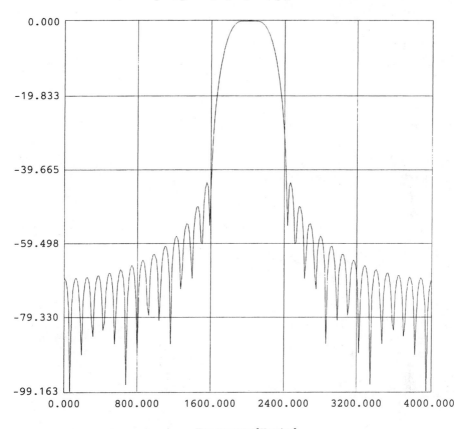

Log Magnitude[dB] Response

Frequency[Hertz]

All Frequencies in Hertz	Digital Filter Design
Sampling Frequency : 8000	
Passband Cutoff Frequency : 1900 2100	Bandpass Filter
Stopband Cutoff Frequency : 1600 2400	FIR Windows
Stopband Ripple in dB : 40	Kaiser Function
	Number of Taps : 61
	Floating point

FIGURE B.7 Magnitude response of FIR bandpass filter using DigiFilter.

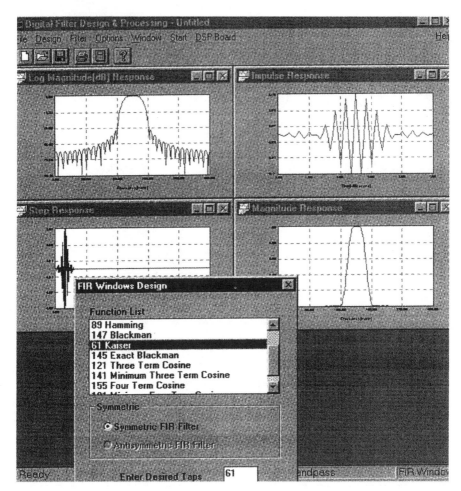

FIGURE B.8 Responses of FIR filter using DigiFilter.

Example B.3 IIR Filter Design and Implementation Using DigiFilter

An IIR filter can be readily designed with the filter package DigiFilter and implemented in real-time on the DSK, without any programming. As with an FIR design, one can choose among several designs using the following functions: Butterworth, Chebyshev, Elliptic, and Bessel, each associated with a specific filter order. A plot of the magnitude response similar to an FIR design, as well as a plot of the poles and zeros of $H(z)$ can be obtained.

B.4 MATLAB FOR FIR/IIR FILTER DESIGN, FFT, AND DATA ACQUISITION

FIR and IIR filters can be designed using the MATLAB software package [6]. FFT and IFFT functions are also available with MATLAB.

Example B.4 FIR Filter Design Using MATLAB

Figure B.9 shows a listing of a MATLAB program MAT33.M to design a 33-coefficient FIR bandpass filter. The function remez uses the Parks–McClellan algorithm based on the Remez exchange algorithm and Chebyshev's approximation theory. The desired filter has a center frequency of 1 kHz with a sampling frequency of 10 kHz. The frequency v represents the normalized frequency variable, defined as $v = f/F_N$, where F_N is the Nyquist frequency. The bandpass filter is represented with 3 bands:

1. The first band (stopband) has normalized frequencies between 0 and 0.1 (0–500 Hz), with corresponding magnitude of 0.
2. The second band (passband) has normalized frequencies between 0.15 and 0.25 (750–1,250 Hz), with corresponding magnitude of 1.
3. The third band (stopband) has normalized frequencies between 0.3 and the Nyquist frequency of 1 (1500–5000 Hz), with corresponding magnitude of 0.

Run this program from MATLAB and verify the magnitude response of the ideal desired filter plotted within MATLAB in Figure B.10. Note that the frequencies 750 and 1250 Hz represent the passband frequencies with normalized frequencies of 0.15 and 0.25, respectively, and associated magnitudes of 1. The frequencies 500 and 1500 Hz represent the stopband frequencies with normalized frequencies of 0.1 and 0.3, respectively, and associated magnitudes of 0.

```
%MAT33.M FIR BANDPASS WITH 33 COEFFICIENTS USING MATLAB Fs=10 kHz
nu=[0 0.1 0.15 0.25 0.3 1];    %normalized frequencies
mag=[0 0 1 1 0 0];             %magnitude at normalized frequencies
c=remez(32,nu,mag);            %invoke remez algorithm for 33 coeff
bp33=c';                       % coeff values transposed
save matbp33.cof bp33 -ascii;  %save in ASCII file with coefficients
[h,w]=freqz(c,1,256);          %frequency response with 256 points
plot(5000*nu,mag,w/pi,abs(h))  %plot ideal magnitude response
```

FIGURE B.9 MATLAB program for FIR filter design (MAT33.M).

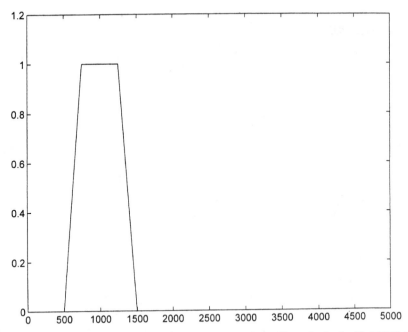

FIGURE B.10 Frequency response of desired FIR bandpass filter obtained with MATLAB.

The last normalized frequency value of 1 corresponds to the Nyquist frequency of 5000 Hz and has a magnitude of zero.

The program generates a set of 33 coefficients saved into the coefficient file matbp33.cof in ASCII format. This file needs to be modified to incorporate the .float assembler directives necessary to implement in C3x code the filter on the DSK. Test your results with the program FIRPRN.ASM, with an internally generated pseudorandom noise as input.

Example B.5 Multiband FIR Filter Design Using MATLAB

This example extends the previous three-band example to a five-band design in order to obtain two passbands. The program MAT63.M shown in Figure B.11 is similar to the previous MATLAB program MAT33.M. This filter with two passbands is represented with a total of five bands: the first band (stopband) has normalized frequencies between 0 and 0.1 (0–500 Hz), with corresponding magnitude of 0; the second band (passband) has normalized frequencies between 0.12 and 0.18 (600–900 Hz), with corresponding magnitude of 1, and so on. This is summarized as follows:

```
%MAT63.M TWO PASSBANDS-FIR WITH 63 COEFFICIENTS USING MATLAB Fs=10 kHz
nu=[0 0.1 0.12 0.18 0.2 0.3 0.32 0.38 0.4 1];  %normalized frequencies
mag=[0 0 1 1 0 0 1 1 0 0];      %magnitude at normalized frequencies
c=remez(62,nu,mag);             %invoke remez algorithm for 63 coeff
bp63=c';                        % coeff values transposed
save mat2bp.cof bp63 -ascii;    %save in ASCII file with coefficients
[h,w]=freqz(c,1,256);           %frequency response with 256 points
plot(5000*nu,mag,w/pi,abs(h))   %plot ideal magnitude response
```

FIGURE B.11 MATLAB program for two-passbands FIR filter design (MAT63.M).

Band	Frequency, Hz	Normalized f/F_N	Magnitude
1	0–500	0–0.1	0
2	600–900	0.12–0.18	1
3	1000–1500	0.2–0.3	0
4	1600–1900	0.32–0.38	1
5	2000–5000	0.4–1	0

Run this program from MATLAB, and verify the magnitude response of the ideal two-passbands filter in Figure B.12. This program generates a set of 63 coefficients saved into the coefficient file mat2bp.cof in ASCII format. Edit this file to incorporate the .float assembler directives necessary to include it in one of the FIR filter programs discussed in Chapter 4. Test your results.

Example B.6 IIR Filter Design Using MATLAB

MATLAB can be used also for the design of IIR filters. A function yulewalk, available in MATLAB, allows for the design of recursive filters based on a best least squares fit [6]. Consider again the MATLAB program MAT33.M in Figure B.9 to obtain a 33-coefficient FIR bandpass filter centered at 1000 Hz. In lieu of the remez function for an FIR design, the following command:

$$\text{>> [a,b] = yulewalk(n,nu,mag)}$$

returns the a and b coefficients in the general input–output equation associated with an IIR filter of order n. The C3x and C programs in Chapter 5 implement an IIR filter with cascaded second-order sections, as is most commonly done. For example, if $n = 6$ in the yulewalk function, the general transfer function in Chapter 5 in terms of the resulting a and b coefficients from MATLAB needs to be reduced to one in terms of three cascaded sections.

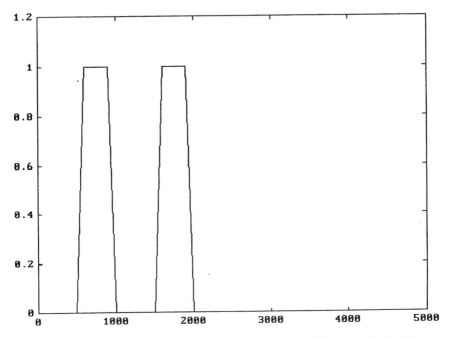

FIGURE B.12 Frequency response of two-passbands FIR filter using MATLAB.

Example B.7 *H*(*z*) from *H*(*s*) Using Bilinear Function in MATLAB

Using Exercise 5.3 in Chapter 5 with the second-order IIR bandstop filter, the following transfer function in the analog *s*-plane

$$H(s) = \frac{s^2 + 0.5271}{s^2 + 0.096s + 0.5271}$$

can be converted to an equivalent transfer function in the digital *z*-plane using the bilinear function from MATLAB with the following commands:

```
>>num = [1,0,0.5271];        %numerator coefficients in s-plane
>>den = [1,0.096,0.5271];    %denominator coefficients in s-plane
>>T = 2; Fs = 1/T;           %K=1 from bilinear equation in Ch 5
>> [a,b]=bilinear(num,den,Fs) %invoke bilinear function
```

to obtain the coefficients *a* and *b* associated with the transfer function in (5.4), or

$$H(z) = \frac{0.9409 - 0.5827z^{-1} + 0.9409z^{-2}}{1 - 0.5827z^{-1} + 0.8817z^{-2}}$$

which is the same transfer function as found in Exercise 5.3 with the bilinear transformation program BLT.BAS, written in BASIC. Note that $T = 2$ was chosen with MATLAB since the constant $K = 2/T$ in the bilinear equation in Chapter 5 was set to 1 for convenience. Note that MATLAB uses the following notation in the general input–output equation:

$$y(n) = b_0 x(n) + b_1 x(n-1) + b_2 x(n-2) + \ldots - a_1 y(n-1) - a_2 y(n-2) - \ldots$$

which yields a transfer function of the form

$$H(z) = \frac{b_0 + b_1 z^{-1} + b_2 z^{-2} + \ldots}{1 + a_1 z^{-1} + a_2 z^{-2} + \ldots}$$

which shows that MATLAB's a and b coefficients are the reverse of the notation used in (5.1).

Example B.8 Eight-Point FFT and IFFT Using MATLAB

The eight-point FFT in Exercise 6.1 can be readily verified with MATLAB, with the following commands:

```
>>x = [1 1 1 1 1 0 0 0 0];
>>y = fft(x)
>>magy = abs(y)
>>plot(magy)
```

The output magnitude is also plotted.

Similarly, the inverse FFT can also be verified.Given the output sequence X's in Exercise 6.1, the inverse FFT or IFFT is:

```
>>X = [4 1-2.414*i 0 1-0.414*i 0 1+0.414*i 0 1+2.414*i];
>>y = ifft(X)
```

where y is the resulting rectangular sequence.

Example B.9 Data Acquisition, Plotting, and FFT With MATLAB

Example 3.13 illustrates the data acquisition of 512 points into a file daq.dat. Figure B.13 shows the MATLAB program DAQ.M that loads and executes the PC host program DAQ.EXE (on disk) with the DOS command !daq. See Example 3.13. This PC host program also downloads and runs on the DSK the executable DSK file DAQ.DSK, and creates 512 data sample points stored in the file daq.dat. The MATLAB DAQ.M program loads the data file daq.dat from MATLAB, and plots the 512 data points in both the time and frequency

```
%DAQ.M - DATA ACQUISITION OF 512 POINTS
while 1,                          %loop continuously
!daq                              %execute data acquisition program
load daq.dat;                     %load file with 512 sample points
plot(daq);                        %plot samples using Matlab
pause                             %wait for a key
R1=fft(daq,512);                  %512-point FFT using Matlab
P1=abs(R1)/512;                   %take magnitude
fs = 44640;                       %Fs same as in DAQ.ASM for DSK
f=[0:fs/512:fs-fs/512];           %frequency axis between 0 - Fs
plot(f(1:256),P1(1:256));         %plot FFT between 0 - Fs/2
title('Frequency Spectrum');      %plot title
pause                             %wait for a key
end                               %loop back
```

FIGURE B.13 MATLAB program for data processing of 512 sample points (DAQ.M).

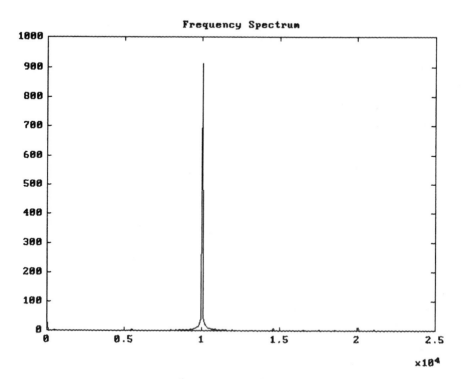

FIGURE B.14 Frequency spectrum of 10-kHz signal with FFT and plot using MATLAB.

domains. The sampling rate is 44.64 kHz, the same as set in the DSK program DAQ.ASM.

Access MATLAB and type DAQ which runs DAQ.M. Input a 10-kHz sinusoidal signal. This time-domain signal will be displayed on the PC monitor screen, plotted from MATLAB. Press any key and the FFT of the 10-kHz signal will be displayed on the PC monitor as shown in Figure B.14.

Change the input signal frequency to 3 kHz. Press any key to obtain the time-domain display of the 3-kHz signal. Deleting the pause commands in the MATLAB program DAQ.M yields continuous displays in both time and frequency domains. Press CTRL-BREAK to stop execution.

Real-Time FIR/IIR Filter Design Using MATLAB Interfaced with the DSK

A program has been developed which interfaces MATLAB directly with the DSK for real-time FIR and IIR filter design [8]. This program makes use of modified versions of FIRNC.ASM and IIR6BPC.C discussed in Chapters 4 and 5. It calculates the TA and TB values for AICSEC, sets the necessary format for the coefficients and the AICSEC data, assembles and loads/runs the appropriate support files.

To test this "interface" program, the MATLAB functions fir2 (for a window design with an arbitrary shape) and remez (for an equiripple design) are used. For an IIR design, the MATLAB functions butter (or cheby1, cheby2, ellip) are used along with tf2zp and zp2sos to find the transfer function's poles and zeroes and the equivalent second-order sections, respectively.

REFERENCES

1. Code Explorer, from GO DSP, at www.go-dsp.com.
2. Goldwave, at www.goldwave.com.
3. Cool Edit, at www.syntrillium.com.
4. Gram412.zip from Spectrogram, address from shareware utility with the data-base address www.simtel.net.
5. DigiFilter, from MultiDSP, at multidsp@aol.com.
6. *The Student Edition of MATLAB.* The Math Works Inc., Prentice Hall, Englewood Cliffs. NJ, 1992.
7. Internet Resource from Rice University, at spib.rice.edu.
8. W. J. Gomes III and R. Chassaing, "Real-Time FIR and IIR Filter Design using MATLAB Interfaced with the TMS320C31 DSK," to be presented at the 1999 ASEE Annual Conference.

C

External and Flash Memory

This appendix describes a homemade daughter board with 32K words (32-bit wide) of external memory and 128K bytes of flash memory. This board connects directly to (fits underneath) the DSK through the four 32-pin connectors JP2-3 and JP5-6 along the edge of the DSK board. All the appropriate signals used on the TMS320C31, such as address, data, V+, GND, R/W, INT0-3, and STRB are available through these four connectors. Figure C.1 shows a photo of the DSK with the daughter board. A specific application program can run from flash without any connection through the PC, as illustrated later. Programming examples test the external memory board with the supporting files on the accompanying disk. Figure C.2 shows a diagram of the daughter board with the external SRAM and flash memory.

CONSTRUCTION OF THE EXTERNAL/FLASH MEMORY BOARD

The external memory board was constructed using a 4″ x 5″ Vector board with copper on one side, with the following parts purchased for approximately $107:

Quantity	Parts description	Price
1	Vector 4″ x 5″ 169P44C1 copper circuit board	$ 16.00
16	3M P/N 3397-1240 wire connection strips	15.00
4	32-pin DIL plug—AMP P/N 1-534206-6	11.00
4	32-pin DIL header—0.1 inch spacing	15.00
1	12-pin DIL header—0.1 inch spacing	2.00
4	28-pin IC socket—0.3 wide	7.00
1	32-pin IC socket—0.5 wide	1.00
2	14-pin IC socket—0.3 wide	1.50
1	2.5mm power connector	3.50
4	14 ns 32K x 8 SRAM	16.00
1	AM29F010-70PC flash memory	11.00

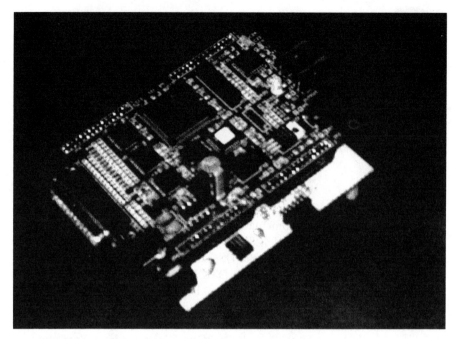

FIGURE C.1 Photo of DSK and daughter board with external and flash memory.

Quantity	Parts description	Price
1	74AS27 3-input NOR gate	1.00
1	74S00 2-input NAND gate	1.00
7	0.1µF ceramic decoupling capacitors	1.00
1	47µF/25V tantalum capacitor	1.00
1	10µF/10V tantalum capacitor	1.00
1	LM340T5 voltage regulator	2.00
6	(3) 10K and (3) 470 ¼ watt resistors	1.00
	Total	$ 107.00

The four 32-pin DIL plugs are to be carefully soldered into the DSK board. The corresponding holes in the vector board are marked for the mating 32-pin header connectors. A pad cutter is used for all signal pins of the four 32-pin header connectors, except the ground pins. Provide a low-impedance power trace to the SRAM's and supporting IC's, with a power strip between two adjacent perforated holes to create an isolated trace (0.2 inch wide) and route power to all the power pins. After the IC sockets are soldered to the vector board, the wire connection strips can then be soldered to the pins on the IC sockets.

A 5-volt regulator supplies power to the external board, since the current

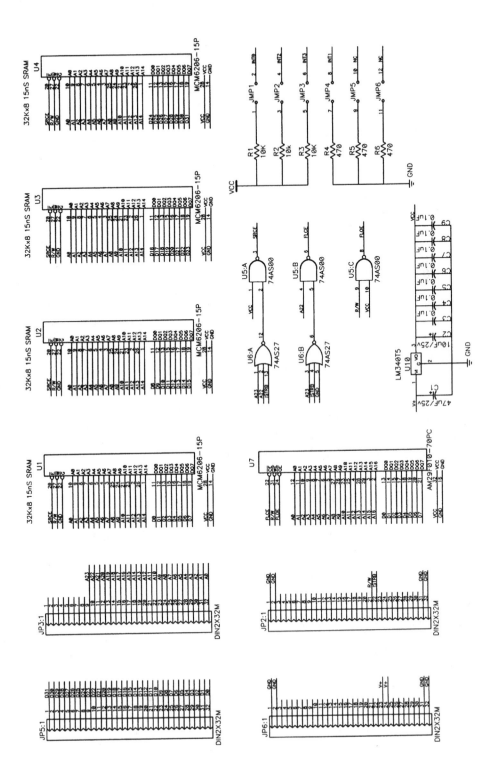

FIGURE C.2 Diagram of external and flash memory.

285

drawn is too high to use the 5-volt supply on the DSK. Connect the power to the V+ pins 23 and 24 from connector JP6 to allow both the DSK and the external board to be powered from one external power supply.

C.1 EXTERNAL MEMORY

The address lines are decoded to provide the following memory decode ranges:

Memory range (Hex)	A22	A23
0–3FFFFF	Low	Low
400000–7FFFFF	High	Low
800000–BFFFFF	Low	High
C00000–FFFFFF	High	High

Since the DSK does not decode any address below 800000h, the first decode range of 0–3FFFFFh is chosen for the external SRAM. The external SRAM can be accessed through any 32K boundary within this range, such as 8000h, 200000h, 300000h, or any valid 8000h boundary within the decoded address range 0–3FFFFF. Since the addresses 0–FFFh are reserved for the boot loader operations on the TMS320C31, as shown in Figure 2.2, the starting address of 100000h is chosen for the external SRAM.

The STRB output is used to qualify for a valid external memory address on the external bus. For example, the STRB line would not be asserted in the memory range reserved for the boot loader operations or associated with any TMS320C31 reserved-memory locations listed in the memory map.

For zero wait state operations, an access time of 15 ns or less for the SRAM chips should be used. A 32K x 8 bit SRAM with these specifications can be purchased, such as the L7C199PC15 from Logical Devices, the MCM6206-15P from Motorola, or the UM61256FK-15. The propagation delay of the logic gates must have a combined delay of less than 6 ns to decode the two MSBs of the TMS320C31 address bus that are used for chip selects of the SRAMs and flash devices.

Testing the External Memory

The program TESTMEM.CPP (on the accompanying disk) is used to test the external memory SRAM. It is to be compiled using Borland's C/C++ compiler. Different data is written to the memory locations 100000h–107FFFh (32K x 32) on the SRAM and the same data is read back. If any memory error is encountered, the program will halt and display the memory location with the error. Execute this program. For example, option 1 writes the value 0xAAAAAAAA to memory, and then reads it back from memory locations 0x100000 to 0x107FFF (32k words).

Example C.1 Multirate Filter With 10 Bands Using External Memory and TMS320C3x Code

The 10-band multirate filter is described in Chapter 8, where a 7-band version was implemented. The 10-band multirate filter requires over 3K words of program and data memory spaces and cannot be implemented on the DSK without additional memory. The program MR10SRAM.ASM (on the accompanying disk) implements the 10-band multirate filter using the daughter board with 32K words of additional memory. Text or code starts at the address 0x100000, which is the starting address of SRAM. The sampling frequency is set at approximately 16 kHz, and band 8 is turned *on*. Download the executable DSK file as done throughout the manual. Verify a bandpass filter centered at 1 kHz. With band 10 *on*, the center frequency is at 4 kHz.

C.2 FLASH MEMORY

The daughter board includes the 1 Megabit or 128K x 8 bit 29F010 flash memory chip, available from Advanced Micro Devices [1–4]. The flash memory address is chosen to start at 400000h to correspond to the second boot region in the TMS320C31 memory map. By configuring INT0, INT2, INT3 High and INT1 Low, the TMS320C31 starts boot loading from address 400000h. Four jumpers, used to obtain these conditions when running from flash, must be removed when downloading a program through external memory to flash.

The flash memory is divided into 8 sectors of 16K Bytes and can be reprogrammed in standard EPROM programmers, requiring a single 5-V power supply operation for both read and write functions. It can store TMS320C3x code and allows the DSK to boot up out of flash memory when the external board is powered. To program the flash, the host PC program downloads a bootable hex file into the external SRAM and the flash memory is then written 8 bits at a time from the external SRAM until the entire hex file is programmed into the flash.

Example C.2 Sine Generation With Four Points From Flash Memory, Using C Code

This example illustrates a sine generation with four points with the program SINEHEX.C shown in Figure C.3, run from flash. All appropriate support files are on the accompanying disk. A version of this sine generation program was tested with the DSK in Experiment 3. The DB25 cable is not connected to the PC, and the program runs when power is applied to the daughter board only. This program runs with zero wait states. By setting the primary bus control register with 0x1018 (see Appendix A), bits 5–7, which designate WTCNT = $(000)_b$ = 0, specify zero wait states [5].

```
/*SINEHEX.C - SINE GENERATION PROGRAM TO RUN FROM FLASH        */
#include "aiccomc.c"                    /*AIC comm routines      */
int AICSEC[4] = {0x162C,0x1,0x4892,0x67};      /*AIC data        */
int data_out, loop = 0;                /*declare global variables */
int sin_table[4]={0,4096,0,-4096}; /*values for 4-point sinewave */

void c_int05()                         /*TINT0 interrupt routine   */
{
 PBASE[0x48]=sin_table[loop] << 2; /*output value from sine table*/
 if (loop < 3) ++loop;                 /*increment loop counter < 3 */
 else loop = 0;                        /*reset loop counter         */
}

main()
{
 unsigned int *pAddr;
 pAddr = (unsigned int *)0x808064; /*primary bus control register*/
 *pAddr = 0x1018;                      /*set WTCNT for 0 wait states */
 pAddr = (unsigned int *)0x808038;
 *pAddr = 0x00100000;                  /*set timer 1 period          */
 pAddr = (unsigned int *)0x808030;
 *pAddr = 0x000003C1;                  /*set timer1 control register */
 AICSET_I();                           /*function to configure AIC   */
 for (;;);                             /*wait for interrupt          */
}
```

FIGURE C.3 Program to generate sine using flash memory (SINEHEX.C).

1. Compile the program SINEHEX.C with the TMS320 tools.
2. Link with the linker command file SINEHEX.CMD, which creates the executable COFF file SINEHEX.OUT and a map file SINEHEX.MAP (both on disk).
3. Locate from the map file the entry address of "_c_int00" as 0x10008C.
4. Edit the file SINHEX30.CMD (on disk) to specify the entry address of 0x10008C.
5. Create the hex file SINEHEX.A0 (on disk) with the command Hex30 SINHEX30.CMD.

6. Download this hex file through the external memory to the flash with the command `C31DLHEX SINEHEX.A0`.

The file `C31DLHEX.CPP` is to be compiled using Borland's C/C++ compiler. The file `HEX30.EXE` is provided with the TMS320 floating-points tools described in Chapter 1. The sine generator program is executed once power is applied to the daughter board, producing a 2-kHz tone. The LED on the DSK board should flash rapidly.

Example C.3 FIR Bandpass Filter From Flash Memory Using C Code

The FIR bandpass filter program `FIRC.C` was implemented in Chapter 4. It is modified to yield `BP45HEX.C` (on disk) to run from flash. Repeat the procedure in the previous example to create and download the hex file `BP45HEX.A0` (on disk) to run from flash. On power to the external board, a bandpass FIR filter with a center frequency at $F_s/10$ or 800 Hz is implemented from flash. Verify these results.

REFERENCES

1. AM29F010 Data Sheet, Publication #16736, Advanced Micro Devices, Sunnyvale, CA, October 1996, at www.amd.com.
2. AM29F010 1 Megabit CMOS 5.0 Volt-only, Sector Erase Flash Memory, Publication #16736, Advanced Micro Devices, Sunnyvale, CA, November 1995.
3. K. Prabhat, Design-in with AMD's AM29F010–Application Note, Publication #17097, Advanced Micro Devices, Sunnyvale, CA, November 1995.
4. K. Prabhat, Reprogrammable Flash BIOS Design Using AMD's AM29F010–Application Note, Publication #17078, Advanced Micro Devices, Sunnyvale, CA, March 1994.
5. TMS320C3x User's Guide, Texas Instruments Inc., Dallas, TX, 1997.

D

Input and Output with 16-Bit Stereo Audio Codec

PETER MARTIN

A homemade wire-wrapped board was designed and built based on the stereo audio codec CS4216 (or CS4218), interfaced to the DSK [1,2]. The CS4216 and CS4218 are functions- and pin-compatible. Figure D.1 shows a diagram of the stereo audio codec CS4218, which provides 16-bit input and output (I/O) alternative with a maximum sampling rate of 50 kHz per channel, and includes two sets of ADCs and DACs. Both ADCs and DACs use delta-sigma modulation with 64x oversampling [3,4]. Antialiasing input and smoothing output digital filters are contained in the codec chip.

An evaluation board, based on the CS4216/CS4218, as shown in Figure D.2, has also been tested and is available from Crystal Semiconductor Corp. [5]. Support programs (on the accompanying disk) can test both the homemade and the evaluation board.

Figure D.3 shows a schematic diagram of the homemade board, which interfaces to the DSK and contains jacks for line as well as microphone inputs. Figure D.3 (a) is the main interface circuit and Figure D.3 (b) and (c) show the input and output circuits, respectively. A header connector on the homemade board connects to JP1 on the DSK board through a ribbon-cable connector. The jumper connectors on JP1 must be removed to disconnect the AIC on the DSK board.

The signal pins SSYNC, SCLK, SDIN, and SDOUT on the codec are used for serial communications and interface with FSX0, FSR0, CLKR0, and DR0 pins on the TMS320C31. The interface includes the codec reset signal that is supplied from XF0 on the TMS320C31. The TCLK0 on the TMS320C31 is used as the codec master clock which controls the ADC and DAC conversion rates. The signal pins on the left of the schematic in Figure D.3 (a) are connected to amplification-control circuits for line-level I/O, microphone input, and speaker output. LIN1 and RIN1 pins can be connected to either a ½-gain

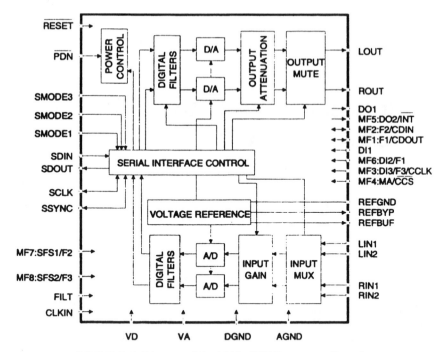

FIGURE D.1 Diagram of the 16-bit CS4218 stereo audio codec.

(regular or line input) or to a 16-gain (microphone input) amplifier using jumpers. The REBUF signal is used for biasing the amplifiers and allow input signals with no DC offset. The LOUT and ROUT signal pins are connected to a coupling circuit (line or regular output) and amplifier to drive a speaker. The SMODE1-3 signal pins allow 32-bit serial port transfers, and MF2-8 allow for the maximum sampling rate and configure the codec-control serial port for default control settings.

The processor's TCLK0 signal is used as the codec master clock, which controls the A/D and D/A conversion rates. With the timer set to pulse mode, different sampling rates can be achieved by setting the appropriate value in the period register. Since

$$F_s = \text{CLKIN}/256$$

the sampling frequency is

$$F_s = (MCLK/4)/(PR{\cdot}256) = (50 \text{ MHz}/4)/(PR{\cdot}256) = 48.828 \text{ kHz}/PR$$

where $MCLK$ is the processor's master clock, and PR is the period register val-

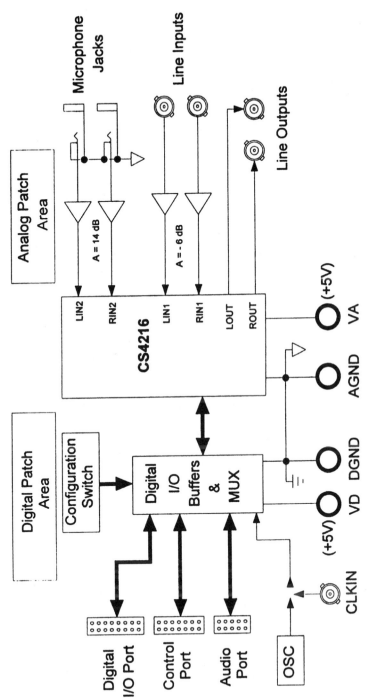

FIGURE D.2 Diagram of the CS4216 Evaluation Board.

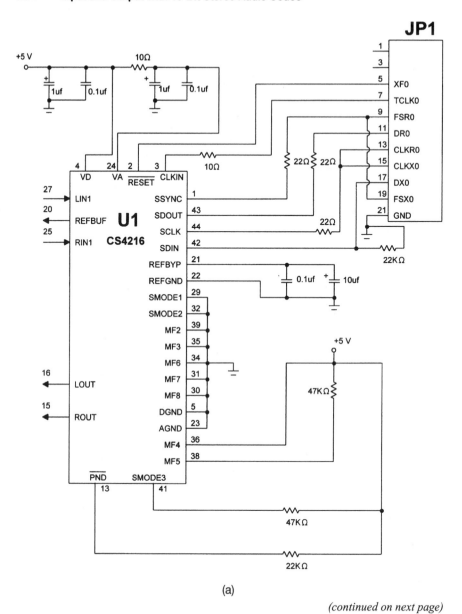

(a)

(continued on next page)

FIGURE D.3 Schematic Diagram of the homemade board using the CS4216 stereo audio codec. (a) interface schematic; (b) input circuits; (c) output circuits.

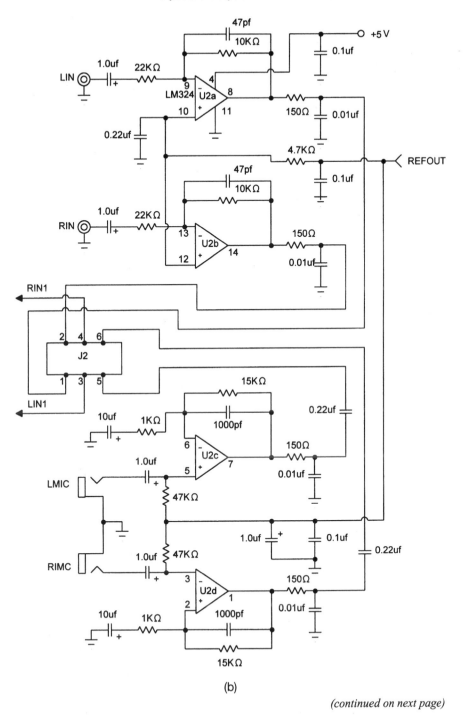

(b)

(continued on next page)

FIGURE D.3 *(continued)*

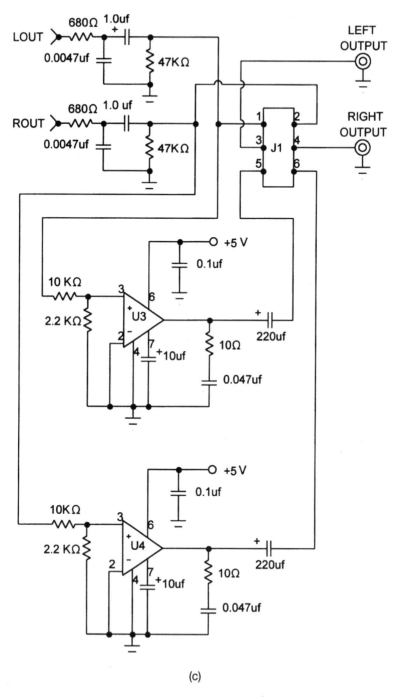

(c)

FIGURE D.3 *(continued)*

ue. For example, the sampling rate is 48,828 Hz by setting the period register with a value of 1. Several sampling frequencies can be programmed as follows:

Period register	Sampling frequency, kHz
1	48.828
2	24.414
3	16.276
4	12.207
5	9.766
6	8.138
⋮	⋮

These rates are illustrated with a programming example of an FIR filter centered at a frequency of $F_s/10$. The following examples can be used to test the Crystal codec.

Example D.1 Loop Programs for Input and Output With the Crystal 16-Bit Stereo Audio Codec Using TMS320C3x Code

Figure D.4 shows a listing of the program LOOPL_CS.ASM that includes the communication routines contained in CSCOM.ASM (on disk). Assemble LOOPL_CS.ASM (not CSCOM.ASM) and verify that an input signal into the

```
;LOOPL_CS.ASM - LOOP PROGRAM USING LEFT CHANNEL OF CS4216/CS4218
        .start   ".text",0x809900 ;where text begins
        .start   ".data",0x809C00 ;where data begins
        .include "CSCOM.ASM"      ;CS codec comm routines
        .text                     ;assemble into text section
        .entry   BEGIN            ;start of code
BEGIN   LDP      SETSP            ;init data page
        LDI      1,R0             ;Fs = 48.8 KHz
        STI      R0,@SRATE        ;Fs in address of CSCOM.ASM
        CALL     COMSET           ;init codec CS/TMS320C31 interface
        LDI      0,R7             ;init output in R7 to zero
LOOP    CALL     IO_L_P           ;left I/O routine,input=>R6,out=>R7
        LDI      R6,R7            ;output R7 = new input sample
        BR       LOOP             ;loop continuously
        .end                      ;end
```

FIGURE D.4 Loop program to test the left channel of the Crystal codec (LOOPL_CS.ASM).

left channel of the Crystal-based board yields from the left output channel a delayed output of the same frequency as the input signal.

A similar program LOOPR_CS.ASM tests the right channel of the Crystal codec, and the program LOOPB_CS.ASM (on the accompanying disk) tests both channels.

Example D.2 FIR Filter With the Crystal Stereo Audio Codec Using TMS320C3x Code

The 45-coefficient FIR bandpass filter illustrated in Chapter 4 is implemented in this example with the Crystal codec with the program BP45CS.ASM (on disk). Access the right I/O channel. Assemble and run this program and verify a bandpass filter centered at $F_s/10$. Substitute different values for the period register. For example the instruction LDI 2,R0 in the source program sets a sampling rate of 24.414 kHz.

REFERENCES

1. CS4218 16-Bit Stereo Audio Codec, Crystal Semiconductor Corp., Austin, TX, September 1996.
2. CS4216 Stereo Audio Codec, Crystal Semiconductor Corp., Austin, TX, October 1993.
3. J. C. Candy and G. C. Temes eds., *Oversampling Delta-Sigma Data Converters—Theory, Design and Simulation,* IEEE Press, New York, 1992.
4. P. M. Aziz, H. V. Sorensen, and J. Van Der Spiegel, "An Overview of Sigma Delta Converters," *IEEE Signal Processing Magazine,* January 1996.
5. CDB 4216 Evaluation Board, Crystal Semiconductor Corp., Austin, TX, June 1993.

Index